Business Letters

用英文打好人際關係

英文 商業書信

一看就會

It's so easy！

序 言

根據人力資源網站的調查，英文能力已成為求職時的基本技能，而能夠寫出具說服力的英文商業書信更是與客戶溝通的必備條件。然而要怎麼寫信給客戶呢？要如何用英文 promote 公司產品呢？別傷腦筋，葵花寶典在此！舉凡商業書信會遇到的情境，本書都有範例供參考，讓讀者使用起來得心應手。

Sama Yavoo求學時留歐，英文系畢業（暗示：這本書的英文很正點），在商業的領域也快十年（經驗十足），舉凡與國外客戶商業書信往來會遇到的狀況，她都能駕輕就熟應付，寫起這本書更是得心應手。從書信的格式、履歷表、開發信、詢價、報價、下單、訂單修改、出貨、付款、客訴、出國參展、謝函等等，本書都有詳盡而實用的範例，讓讀者與國外客戶書信往來時更具說服力，進而在職場上一展長才。

想要在競爭激烈的職場中佔有一席之地嗎？對國外客戶的書信會不知所措嗎？只要擁有本書，一切的問題都將迎刃而解！它將是你求職、開發客戶、生意往來的最佳工具。

Sama Yavoo與我相交多年，能夠幫她及她的團隊寫序是我的榮幸，像她這麼認真的人，寫的書也一樣認真和詳盡，所以我可以很放心的大膽推荐此書給各位。

鄭惠明／高明輝

致讀者

　　浸淫在英文商業書信領域中已有近十年歲月,英文有句諺語是這麼說的:"Practice makes perfect."中文意思是藉由不斷的練習可以趨近完美。其實「時間」對這句諺語扮演舉足輕重的地位,因為透過時間的不斷練習,自然能熟能生巧。

　　透過本書的撰寫,使我對過去十年來在學校和工作上所學的一切,有一個回顧和審視的機會,也算是對英文商業書信做了一個很完整的複習和總結。

　　隨著時代不斷改變,英文商業書信的表達方式,也從傳統式的書信到傳真,再到上一個世紀末出現的e-mail,但其格式與內容本身的改變卻不是很大。商業書信注重言簡意賅,不論是書信、傳真或e-mail,它的中心思想是不變的。

　　本書並未針對傳真或e-mail加以著墨,主因是英文商業書信是最基本的格式,而傳真和e-mail則是英文商業書信的變體。整體說來,傳真省略了收信人的地址,但其他的格式則與英文商業書信大同小異;而e-mail則是將對方和自己的地址改成電子郵件信箱,其他的格式與內容則與英文商業書信相似。

　　本書對商業書信的構成要素、格式和注意事項做了清楚的說明。另外，在編排順序上，則是從應徵工作為始，之後則是以完整的商業貿易流程為主，按順序、分階段來寫成。透過本書，讀者除了可習得如何寫一封專業的英文書信外，更可從中獲得與貿易流程相關的知識，非常適合那些已具英文基礎的人做商業入門書，也適合相關人士當做參考教材。

英文商業書信的書寫並不困難，本書以淺顯易懂的語言、詳細的用詞說明，讓讀者閱讀起來有事半功倍的效果。書中所提的實例，在商業書信中皆是非常常見和實用的例子。希望讀者能自本書中受惠。

切記！本書雖已提供許多實用的範例，但用意是在拋磚引玉，希望讀者能據此寫出更適合自己行業、更符合自己狀況的專業英文書信。

CONTENTS

5 詢價、報價、議價

(Pricing Inquiry, Quotation, Counter Offer)   122

The Formates of Business Letter

商業書信之格式

　　工欲善其事，必先利其器。若要寫好英文商業書信也必須對其書寫格式和其構成要素有一定的了解。了解英文商業書信的格式與構成要素有助於寫出一份具專業性和讓人一目瞭然的書信。

1-1 書寫格式

　　英文商業書信最常見的格式有三種：齊頭式 (full-block form / block form)、半齊頭式 (semi-block form)和縮格式 (indented form)。

齊頭式

信頭 (Letterhead)

日期 (Date)
收信人地址 (Addressee's address/Inside Address)

收件人姓名 (Addressee/Attention Line)

稱謂語 (Salutation)

主旨 (Subject Line)

信文 (Body of Letter)

結尾敬辭 (Complimentary Close)
簽名欄 (Signature Block)
識別記號 (Identification Marks)
附件註明 (Enclosure Notation)
副本註明 (Carbon Copy Notation)
附註 (Postscript)

所有的段落除了信頭置中外,其餘的一律從最左邊開始書寫,如上例所示。

半齊頭式

信頭 (Letterhead)

日期 (Date)

收信人地址 (Addressee's address/Inside Address)

收件人姓名 (Addressee/Attention Line)

稱謂語 (Salutation)

主旨 (Subject Line)

信文 (Body of Letter)

結尾敬辭 (Complimentary Close)
簽名欄 (Signature Block)
職稱 (Title)

識別記號 (Identification Marks)
附件註明 (Enclosure Notation)
副本註明 (Carbon Copy Notation)
附註 (Postscript)

信頭 (Letterhead) 置中，日期 (Date)、結尾敬辭 (Complimentary Close)、簽名欄 (Signature Block)、職稱 (Title Line) 向中間齊頭書寫，其餘一律由最左邊齊頭開始書寫。

縮格式

信頭 (Letterhead)

日期 (Date)

收信人地址 (Addressee's Address/Inside Address)

收件人姓名 (Addressee/Attention Line)

稱謂語 (Salutation)

主旨 (Subject Line)

　　信文 (Body of Letter)

　　　　　　　　結尾敬辭 (Complimentary Close)
　　　　　　　　簽名欄 (Signature Block)

識別記號 (Identification Marks)
附件註明 (Enclosure Notation)

　　縮格式與半齊頭式的相同，即除了信頭 (Letterhead) 置中，日期 (Date)、結尾敬辭 (Complimentary Close)、簽名欄 (Signature Block)、職稱 (Title Line) 向中間齊頭書寫外，其餘一律由最左邊齊頭開始書寫。

注意

　　信文部份，每一段落的起頭都要往內縮格，一般是往內縮五格，如上例所示。

1·2 構成要素

　　要寫好一份專業英文商業書信，只知道商業書信的書寫格式還不夠，也必須要知道一封商業書信到底是由哪些基本要素所構成。商業書信構成的要素有：信頭、日期、收信人地址、收信人姓名、稱謂語、主旨、信文、結尾敬辭、簽名、識別記號、附件註明、副本註明和附註。

a. 信頭 (Letterhead)

　　信頭包括寄信人的公司名稱、地址、電話、傳真號碼等基本資料。大部分的公司也會將其商標或標識也放入信頭中。現在更因為 E 化時代的來臨，越來越多的公司也將其網站、電子郵件信箱等資訊放入信頭中。總之，信頭所代表的不僅是一家公司的基本聯絡資料，更包含了一家公司的整體形象。通常信頭都出現在信紙最上方的位置，而且是置中對齊，如下例所示：

〿 American Chemical Co., Ltd. 〿

3F, No. 600 Stafford Avenue, Phoenix, AZ, 72117, U.S.A.

TEL: 501 21 23 56 57 58 FAX: 501 44 21 23 56 57 59

http://www.Americanche.com/ e-mail: Amemical@online.net

Date:

Addressee's Address:

b. 日期 (Date)

　　日期的寫法有美式與歐式兩種：美式寫法的順序是月、日、年，如May 8th, 2013 ；歐式寫法的順序為日、月、年，如 8th May, 2013。

c. 收信人地址 (Addressee's Address/Inside Address)

　　收信人地址的寫法與信頭上地址的寫法不同。信頭上地址的寫法是行列式；而收信人地址的寫法則為區塊式。共分成以下區塊來寫：公司名稱、街道名稱、市、州、郵遞區號、國名。

📎 American Chemical Co., Ltd. 📎

3F, No. 600 Stafford Avenue, Phoenix, AZ, 72117, U.S.A.

TEL: 501 21 23 56 57 58 FAX: 501 44 21 23 56 57 59

http://www.Americanche.com/ e-mail: Amemical@online.net

Date: Nov. 22th, 2013

Jack's Water Treatment Inc.

1890 Dickenson Drive

Austin Texas 79803

USA

　　地址的寫法有兩種，其一為不加上任何標點符號來區別街道名稱、市、州、國等；其二為加上標點符號來區別街道名稱、市、州、國等。故以上的地址可寫成如下：

Jack's Water Treatment Inc.

1890 Dickenson Drive,

Austin, Texas- 79803,

U.S.A.

d. 收信人姓名 (Addressee/Attention)

收信人姓名可以是特定的人或部門。若是特定的人的話，通常也會加上此人的頭銜以示尊重或引起注意，寫法有下列幾種：

(1) Att: Mr. Jose Salvador

General Manager

(2) ATT: MR. JOSE SALVADOR , GENERAL MANAGER

(3) Att: Mr. Jose Salvador/General Manager

(4) ATT: MARKETING DEPARTMENT

(5) Att: Manager of Financial Department

注 意

1. 「ATT」之後收信人的每個字母都需大寫；而「Att」之後收信人的字母只需字首大寫即可，如上列所示。

2. 收信人姓名與地址的順序可以互換。換言之可以先寫收信人姓名，再寫收信人地址；也可先寫收信人地址，再寫收信人姓名。

e. 稱謂語 (Salutation)

稱謂語表示對收信人的尊稱，有以下表達方式：

(1) Dear Mr. Richardson,： Mr. 用來表示對男士的尊稱。

(2) Respectful Madame Dorothy,：也可用 Madame 來表示對已婚女士的尊稱。Respectful是非常正式的用法，用來突顯對方德高望重的身分或對對方表示極度的推崇。

(3) Dear Miss Anderson,：用 Miss 來表示對未婚女士的尊稱。

(4) Respectful Ms. Erickson,：當不知對方是否為已婚或未婚的女性時，則用 Ms. 來稱呼，以避免任何尷尬的情況發生。

(5) Dear Mary,：當雙方相當熟悉或親密時，也可直接稱呼對方名稱。

(6) Hi, Jack,：非常口語的稱呼，通常是在雙方很熟悉的情況才使用。

(7) Jefferson,：在雙方很熟悉的情況下，也可直接稱呼對方名字。

(8) Dear Sir/Gentleman,：在不知道對方的姓名時，可直接以 Dear Sir/Gentleman 來稱呼個人。

(9) Dear Sirs/Gentlemen,：在不知道對方的姓名時，可直接以Dear Sirs/Gentlemen來稱呼法人。通常這種情況的收件人是公司或部門。

(10) To Whom It May Concern,：即中文之「敬啟者」或「貴寶號鈞鑒」，通常是不知道收件者是誰，或是寫給公司或部門時才使用到。

注 意

不可在 Mr./Mrs./Ms./Miss/Madame 之後連名帶姓的稱呼；也不可在 Mr./Mrs./Ms./Miss/Madame 之後直接稱呼名字，如 Mr. Jack Johnson 或 Madame Kelly，這是非常不禮貌且會貽笑大方的寫法。

f. 主旨 (Subject Line)

主旨是一封信的主題，讀者見到此主旨時，便可知道此封信的主要內容為何。主旨要簡潔，使讀者可一目瞭然。可用「Sub:」或「Re:」來引導主旨。例如：

(1) **Sub: Your Letter of Sept. 1st, 2013**

（主旨：您 2013 年 9 月 1 日的來信）

(2) **SUB: YOUR LETTER OF SEPT. 1ST., 2013**

（主旨：您 2013 年 9 月 1 日的來信）

(3) **Subject: Your Order No. 91006**

（主旨：貴公司訂單單號 91006）

(4) **SUBJECT: YOUR ORDER NO. 91006**

（主旨：貴公司訂單單號 91006）

(5) Re: L/C Amendment（主旨：修改 L /C）

(6) RE: L/C AMENDMENT（主旨：修改 L /C）

如上例所示，為吸引讀者的注意，有時會將主旨以粗體標示，甚至加上底線以強調。

注意

SUB/RE/SUBJECT 等之後之主旨，每個字母都需大寫；而 Sub/Subject/Re 等之後的主旨，只需字首大寫即可。

g. 信文 (Body of Letter)

信文一定要段落分明，各段落間通常需空雙行間隔。一般而言，信文共分三段。第一段為 Reference Paragraph，說明寫這封信的原因為何，具有承先啟後的作用，讓收信人明白這封信主要是針對哪件事情回覆；第二段說明本信文之主要內容；第三段則表態期待對方的回覆或其他訴求。如下例說明：

Thank you for your letter of Aug. 14th with regards to the L/C amendment. After checking the details, I would like to inform you of the following:

We found out that the bank had typing errors on the latest date of shipment and the total amount. We have asked the bank to revise the L/C as per request. You should be able to get the amended one within next week.

Should you have any further question, please feel free to let me know. By the way, please confirm if you have received the sample sent last week.

❖ 譯文

謝謝您八月十四日有關修改 L/C 的來函。在核對細節後，我想做以下的陳述：

我們發現銀行打錯了最後裝船日和總金額部分。敝公司已請銀行依照您的要求修正。您應可在下週內收到修改後的 L/C。

若還有任何問題，請讓我知道。不知您是否已收到了上週所寄出的樣品？收到後請來信確認。

注 意

若是齊頭式或半齊頭式的信件格式時，信文的每一段都要左齊定邊書寫；若是縮格式的信件時，信文中每一段首都應向內縮三至五格書寫。

h. 結尾敬辭 (Complimentary Close)

結尾敬辭相當於中文之敬啟、謹啟等。以下是常用的結尾敬辭：

(1) 最正式（表尊重或對方德高望重）　　Respectfully,
Respectfully yours,

(2) 正式	Very respectfully yours, Yours truly, Very truly yours, Yours sincerely, Sincerely yours, Very sincerely yours,
(3) 表示親切	Sincerely, Cordially, Cordially yours, Best regards, Warmest regards, Regards, Best wishes,

i. 簽名 (Signature)

發信人除了需親筆簽名外，還需在簽名處下方打上姓名、抬頭、公司名稱或部門名稱。如下例所示：

Ms. Amanda Thomson
General Manager　　　　或
ABC Industrial Co., Ltd.

Ms. Amanda Thomson/Director
Marketing Center

j. 識別記號 (Identification Marks/Reference Initials)

　　用來表示文件的責任所在時，需將寫信者和打字者的名字特別標示出來。**左邊為寫信者，右邊為打字者。不需將全名打上，只需將姓名的首字縮寫即可。**例如寫信者為 David Lee，打字者為 Catherine Rosemary 時，就可做以下的表示：

DL/CR；DL: cr；DL-CR；也可只打上打字員的姓名縮寫，如 cr 等。

k. 附件註明 (Enclosure Notation)

　　當信件內附其他文件時，須在信中註明，例如：

Enclosure: Agreement 　　（附件：同意書）
Enclosure: 2 　　　　　　（附件：兩份）
Encl: Contract 　　　　　（附件：合約書）
Encl: 1 　　　　　　　　（附件：一份）

l. 副本註明 (Carbon Copy Notation)

　　當此信件需寄給或呈給第三者時，需將第三者的名字和抬頭也一併列上。例如這封信除了收件者外，另外要提供給 Mary Jefferson 和 Jerry Wang 時，便可如以下標示：

cc: Mary Jefferson/CEO

CC: Jerry Wang/President

cc: M.J.

CC: J.W.

m. 附註 (Postscript)

　　附註即大家非常熟悉的「PS.」。原先的附註是針對正文有遺漏的部分做補述，可是正式信函最好不要有任何遺漏，而是將有關正文的部分寫在信文中。但是越來越多的趨勢是將「PS.」用在特意給對方強烈印象或要吸引對方的注意。例如："PS. We got your payment of US\$5,000 for your order 1225. Thank you!"（附註：我們已收到貴公司訂單單號1225，美金五千元的款項，謝謝！）。

1·3 版面編排

　　知道商業書信的格式與構成要素後，最重要的是要將書信格式與基本構成要素做恰當與正確的應用。現在針對商業書信的標準編排方式舉例說明。

齊頭式

〔U American Chemical Co., Ltd. 〔U

3F, No. 600 Stafford Avenue, Phoenix, AZ, 72117, U.S.A.

TEL: 501 21 23 56 57 58 FAX: 501 44 21 23 56 57 59

http://www.Americanche.com/ e-mail: Amemical@online.net

（空 2-3 行）

Date: Nov. 22th, 2013

（空 2-3 行）

Jack's Water Treatment Inc.

1890 Dickenson Drive

Austin Texas 79803

USA

（空兩行）

Att: Mr. Jose Salvador/General Manager

（空一行）

Dear Mr. Salvador,

（空一行）

Sub: L/C Amendment

（空一行）

Thank you for your letter of Aug. 14th with regards to the L/C amendment. After checking the details, I would like to inform you of the following:

（空兩行）

We found out that the bank had typing errors on the latest date of shipment and the total amount. We have asked the bank to revise the L/C as per request. You should be able to get the amended one within next week.

（空兩行）

Should you have any further question, please feel free to let me know. By the way, please confirm if you have received the sample sent last week.

（空一行）

Sincerely yours,

Rose Ryian

（空 2-3 行）

Ross Ryian/Manager

（空一行）

RR: am

（空 1~2 行）

Encl: Amended L/c

（空 1~2 行）

CC: Richard Ho/President

縮格式

〔U American Chemical Co., Ltd. 〔U

3F, No. 600 Stafford Avenue, Phoenix, AZ, 72117, U.S.A.

TEL: 501 21 23 56 57 58 FAX: 501 44 21 23 56 57 59

http://www.Americanche.com/ e-mail: Amemical@online.net

（空 2-3 行）

Date: Nov. 22th, 2013

（空 2-3 行）

Jack's Water Treatment Inc.

1890 Dickenson Drive

Austin Texas 79803

USA

（空兩行）

Att: Mr. Jose Salvador/General Manager

（空一行）

Dear Mr. Salvador,

（空一行）

Sub: L/C Amendment

（空一行）

（往內縮 3~5 格）Thank you for your letter of Aug. 14th with regards to the L/C amendment. After checking the details, I would like to inform you of the following:

（空兩行）

（往內縮 3~5 格）We found out that the bank had typing errors on the latest date of shipment and the total amount. We have asked the bank to revise the L/C as per request. You should be able to get the amended one within next week.

（空兩行）

（往內縮 3~5 格）Should you have any further question, please feel free to let me know. By the way, please confirm if you have received the sample sent last week.

（空 1~2 行）

                        Sincerely yours,

                        Rose Ryian

                        （空 2-3 行）

                        Ross Ryian

                        Manager

（空 1 行）

CC: Richard Ho/President

半齊頭式

〖U〗American Chemical Co., Ltd. 〖U〗

3F, No. 600 Stafford Avenue, Phoenix, AZ, 72117, U.S.A.

TEL: 501 21 23 56 57 58 FAX: 501 44 21 23 56 57 59

http://www.Americanche.com/ e-mail: Amemical@online.net

（空 2-3 行）

Date: Nov. 22th, 2013

（空 2-3 行）

Jack's Water Treatment Inc.

1890 Dickenson Drive

Austin Texas 79803

USA

（空兩行）

Att: Mr. Jose Salvador/General Manager

（空一行）

Dear Mr. Salvador,

（空一行）

Sub: L/C Amendment

（空一行）

Thank you for your letter of Aug. 14th with regards to the L/C amendment. After checking the details, I would like to inform you of the following:

（空兩行）

We found out that the bank had typing errors on the latest date of shipment and the total amount. We have asked the bank to revise the L/C as per request. You should be able to get the amended one within next week.

（空兩行）

Should you have any further question, please feel free to let me know. By the way, please confirm if you have received the sample sent last week.

（空一行）

                              Sincerely yours,

                              Rose Ryian
                              （空 2-3 行）
                              Ross Ryian, Manager

（空 2 行）

PS. We got your payment of US$5,000 for your order #2227. Thank you.

注 意

　　上述的英文書信格式是最基本的格式規範。但是，為了書寫版面上的美觀考量，各段落間的空隔行數，並非一定得按照上述的基本格式。書寫者為求版面上的美觀，有時各段間可空一行或數行。因為商業書信除了要注重內容外，也要注重外型——也就是版面的編排。在編排時應注意以下各點：

1. 整篇書信要位於信紙中間位置。不可偏上、偏下、偏左或偏右。必須考慮到讓讀者一目瞭然，視覺上感到舒適。

2. 段間一定要段落分明，不可以將各區塊擠在一起。否則閱讀時不但費力，反而造成視覺上的壓力，讓人一點繼續讀下去的興致也沒有。

3. 如果一頁寫不完，續寫第二頁時，信頭可以保留或省略，但收信人地址、姓名、稱謂語、主旨等都要省略。只要標明日期，然後繼續上文即可。

　　當然，這時更要注意版面，盡量讓版面保持於信紙中間的位置。

1·4
商業書信目的與 5C 原則

a. 商業書信的目的：

商業書信的目的不外乎以下幾類：

1. **告知訊息 (Informing/Advising)**：這類信函的主要目的是告知對方相關的訊息，對方可依據此通知進行下一步行動。如應徵函、詢價、回覆應徵函、回覆詢價、請求降價、裝船通知、包裝通知、款項逾期通知、付款通知、拜訪通知、客訴通知、下訂單、訂單取消通知、無法如期交貨通知等，都是這一類的信函。

2. **說服對方 (Convincing/Persuading)**：這類信函除了告知對方相關訊息外，更重要的是要說服對方接受自己的意見行事。如應徵函、推銷信、要求降價、拒絕客戶要求的信函等，都是這一類信函。

3. **表達善意 (Expressing good wishes)**：這類信函的主要目的是為了保持雙方良好的關係而互相表達善意。如感謝函、慰問函等都是這類的信函。

b. 商業書信的 5C 原則：

時間就是金錢，商業信函不像一般私人信函可以拉拉雜雜的高談闊論、甚至一寫就寫了好幾頁。以下五個原則是撰寫商業信函時必須注意的，由於這五個原則的英文都是以 C 為開頭，故簡稱5C原則：

1. **Correct**（正確性）：小自英文拼字、英文文法，大至書信格式和內容都要正確。

2. **Clear**（清楚明瞭）：書信的內容和語法要讓讀者能清楚明瞭書寫者所要表達的意思。盡量不要使用語意不明的句式結構或複雜的句型。這樣不但浪費時間，又易使讀者混淆和猜測。若讀者猜錯意思，反而適得其反。

3. **Concise**（簡潔有力）：內容要簡潔有力，用簡短的內容來表達重點要義。忙碌的現代人是無暇且無耐心看完一篇長篇大論，言簡意賅的文章內容才是繁忙的商業人士所鍾愛的，畢竟時間就是金錢！

4. **Considerate**（從對方的角度來書寫）：寫信時，要站在對方的立場來思考。也就是說要思考這封信要怎麼寫才能讓對方明白自己所要表達的意思？怎樣寫比較不會冒犯對方？如何表達對方較能接受？請注意，寫信的目的在於溝通，而不是製造更多的問題和誤解。

5. **Courteous**（禮貌性）：禮貌在商業往來是非常重要的一環。禮貌的信函可以塑造公司的良好形象。所以在遣辭用字時，盡量使用較正面的表達方式和語氣。即使是對方不對，也不要使用責備的語氣。總之，在關鍵時刻，有禮貌的信有助於保持良好的關係。

切記時間就是金錢，謹守以上原則，必可以協助你在撰寫商業書信時無往不利。

Writing a Curriculum Vitae

撰寫履歷表

　　你或許是職場老手，抑或是初入社會的新鮮人。有些頗具吸引力的工作機會，常會要求應徵者附上英文履歷。許多應徵者一聽到英文履歷表就退避三舍，眼睜睜看著大好的工作機會拱手讓人。其實，撰寫一份具專業水準的履歷表一點也不困難！你現在或許對這樣的話感到懷疑，我們也完全能了解你內心的疑惑。但是在你看完這一章後，你會認同我們講的這句：「其實，撰寫一份具專業水準的履歷表真的一點也不困難。」

　　一份正式的英文履歷表是由以下三個部分所組成：履歷表、自傳、求職信。以下就針對這三部分來做詳細說明。

2·1

履歷表

　　履歷表英文稱做 curriculum vitae 或 resume。一般而言，履歷表是由以下幾個部分所組合而成：

a. 個人資料 (Personal Data/Personal Information)

其內容應包括應徵者的以下資料：

1. 姓名 (Name)
2. 住址 (Address)
3. 聯絡電話 (Tel)
4. 生日 (Date of Birth)
5. 性別 (Gender; Sex)
6. 年齡 (Age)
7. 婚姻狀況 (Marital Status)
8. 身高 (Height)
9. 體重 (Weight)
10. 出生地 (Birth Place)
11. 血型 (Blood Type)
12. 身體狀況 (Health)
13. 宗教信仰 (Religion)

b. 學歷 (Education)

應徵者可選擇倒敘法或順敘法來陳述與自己相關的學歷背景。所謂倒敘法是應徵者將自己的求學歷程，依反時針方向倒推回去，也就是把最後的學歷寫在第一個；而順序法是應徵者將自己的求學歷程以順時針方向來呈現，也就是把最後的學歷寫在最後一個。你可以依照上述方式，選擇自己喜愛的方式來陳述你的學歷背景。

c. 工作經驗 (Work Experience)

　　應徵者一樣也可依自己喜好的方式，選擇倒敘法或順序法來陳述與自己相關的工作經驗。但是切記，如果學歷選擇倒敘法，那工作經驗的敘述法就須選擇倒敘法，這樣才會一致，不致使人混淆。

d. 其他

　　這部分的範圍包括了最常見的語言能力 (Language Ability)、電腦技能 (Computer Skills)、特殊訓練 (Special Training)、興趣 (Hobbies)、軍役狀況 (Military Service)、希望待遇 (Expected Salary)、可上班日 (Date Available)……亦可再加上你所想要展現的自我特點。

　　基本上，一份專業的履歷表內容需包含以上所提的四點。至於編排方式，則可以依自己的喜好來表現。總之，一份成功的履歷表應有助於主考官對你留下深刻的印象。

　　以下舉三個履歷表的範例作為參考：一份是依倒敘法來敘述；另一份則以順序法來陳述；還有一份是特別針對社會新鮮人所設計的履歷表。

履歷表㈠倒敘法

Curriculum Vitae

Personal Data:

Name: Richard Lee   Date of Birth: June 11, 1976

Address: No. 666, Ming Shen East Road, Taipei, Taiwan, R.O. C.

Tel: (02) 2772-2838   Mobil: 0911-111-111

E-mail address: Richard@ms31.hinet.net

Marital Status: Married   Age: 37   Gender: Male

Birthplace: Taipei City   Blood type: O   Religion: Catholic

Height: 172cm   Weight: 80kgs   Health: Excellent

Education:

1999-2001   National Taiwan University, Master Degree of Business
             Administration

1995-1999   National Cheng Chi University, Bachelor of
             International Trade

1992-1995   Taipei First Boys School (Cheng Kuo Senior High
             School)

Working Experiences:

2007-2013   General Electronic Inc.: Director

2004-2007   Magic Electronics Co., Ltd: Sales Manager

2001-2004   Power Electronics Co., Ltd: Special Assistant to G.M.

Language Ability: Good command in both Chinese & English

French: Can speak, write and understand daily

conversation

Computer Skills: Win 7, Word, Excel, Lotus, Basic

Special Training: Auditor of ISO/Licensee of Win 7

Hobbies: Playing tennis and piano, reading and traveling

Military Service: ROC Air Force

Expected Salary: NT$60,000/month

Date Available: Since June 1st

❖ 譯文

~~~~~~

履歷表

個人資料：

姓名：Richard Lee　　　　　　　生日：民國 65 年 6 月 11 日
地址：臺北市民生東路 666 號　　手機：0911-111-111
電話：(02) 2772-2838　　　　　　E-mail：Richard@ms31.hinet.net
婚姻狀況：已婚　　　年齡：37 歲　　　性別：男
出生地：臺北市　　　血型：O 型　　　宗教：天主教
身高：172cm　　　　體重：80kgs　　　健康：佳

教育背景：
1999-2001　　國立臺灣大學企管碩士班畢業
1995-1999　　國立政治大學國貿系畢業
1992-1995　　臺北建國中學畢業

工作經驗：
2007-2013　　大眾電機公司：處長
2004-2007　　魔力電機公司：業務經理
2001-2004　　動力電機公司：總經理特助

語言能力：中、英文流利
　　　　　　法文：可以說、寫；瞭解日常會話

電腦技能：Win 7, Word, Excel, Lotus, Basic

特殊訓練或證照：ISO 稽核員證照／Win 7 工程師執照

興趣：網球、鋼琴、閱讀、旅行

軍役：臺灣空軍

希望待遇：NT$60,000／月

可上班日：6 月 1 日起

~~~~~~

履歷表㈡順序法

Curriculum Vitae

Personal Data:

Name:Miranda Huang   Date of Birth: Sept. 1st, 1981

Address: No. 666, Fu Shin North Rd., Taipei, Taiwan, R.O. C.

Tel: (02) 2772-2838   E-mail address: mirandah@ms31.hinet.net

Marital Status: Single   Age: 32   Gender: Female

Birthplace: Taipei City   Blood type: A   Religion: Buddhist

Height: 165 cm   Weight: 55 kgs   Health: Excellent

Education:

1995-1998   Taipei First Girls Shool

1998-2002   National Cheng Chi University, Bachelor of Financing

2002-2004   National Tsing Hua University, Master Degree of
              Accounting

Working Experiences:

1998-2003   The First Law Office: Assistant Accountant

2006-2009   Price Water House: CPA

2009-2013   Coopers Inc. Managing Director

Language Ability: Egnlish: excellent in listening, speaking, reading
              and writing.

Chinese: excellent in listening, speaking, reading and writing

German: good in listening, speaking, reading, and writing

Computer Skills: Basic, Fortran, Win 7, Word, Excel, Lotus

Special Training or Certificate: CPA

Hobbies: Golf, violin, reading and swimming

Expected Salary: NT$120,000/month

Date Available: Since June 1st

❖ 譯文

〰〰〰〰〰〰〰〰〰〰〰〰〰〰〰〰〰〰〰〰〰〰〰〰〰〰

履歷表

個人資料：

姓名：Miranda Huang　　　　　生日：民國 70 年 9 月 1 日
地址：臺北市復興北路 666 號
電話：(02) 2772-2838　　　　　E-mail：mirandah@ms31.hinet.net
婚姻狀況：單身　　年齡：32 歲　　　　性別：女
出生地：臺北市　　血型：A 型　　　　宗教：佛教
身高：165 cm　　　體重：55 kgs　　　健康：佳

教育背景：
1995-1998　北一女畢業
1998-2002　國立政治大學經濟系畢
2002-2004　國立清華大學會計碩士畢

工作經驗：
2004-2006　第一法律事務所：助理會計
2006-2009　資誠會計：會計師
2009-2013　庫柏會計：處長

語言能力：英文：　聽、說、讀、寫均優異
　　　　　中文：　聽、說、讀、寫均優異
　　　　　德文：　聽、說、讀、寫均良好

電腦技能：Basic, Fortran, Win 7, Word, Excel, Lotus,

特殊訓練或證照：會計師執照

興趣：高爾夫球，小提琴，閱讀，游泳

希望待遇：NT$120,000 ／月

可上班日：六月一日起

〰〰〰〰〰〰〰〰〰〰〰〰〰〰〰〰〰〰〰〰〰〰〰〰〰〰

履歷表㈢社會新鮮人

Resume

Personal Data:

Name: Jessica Fang   Date of Birth: May 8th, 1987

Address: No. 666, Shan Ming Rd., Shan Ming District, Kaohsiung, Taiwan, R.O. C.

Tel: (07)7772-2838   E-mail address: jessicaf@ms54.hinet.net

Marital Status: Single   Age: 26   Gender: Female

Birthplace: Kaohsiung City   Blood type: B   Religion: Taoism

Height: 160 cm   Weight: 50 kgs   Health: Excellent

Education:

2011-2013   National Tsing Hua University, Master Degree of
English Linguistics

2009-2011   National Cheng Chi University, English Department

2005-2009   Wen Tzao Foreign Language Junior College
Major: English/Minor: Spanish

Experiences:

2011-2013   Part time Translator

2009-2011   English Tutor

2007-2008   Head of Association of English Department

2006-2007   Head of Student Union of Wen Tzao

Language Ability: Egnlish: excellent in listening, speaking, reading and writing

Chinese: excellent in listening, speaking, reading and writing

Spanish: good in listening, speaking, reading, and writing

Computer Skills: Win 7, Word, Excel, Lotus

Hobbies: movies, music

Expected Salary: NT$30,000/month

Date Available: Since July 1st

❖ 譯文

〜〜

履歷表

個人資料：

姓名：Jessica Fang　　　　　　　　生日：民國 76 年 5 月 8 日
地址：高雄市三民區三民路 666 號
電話：(07) 7772-2838　　　　　　　E-mail：jessicaf@ms54.hinet.net
婚姻狀況：單身　　　年齡：26　　　　　性別：女
出生地：高雄市　　　血型：B　　　　　宗教：道教
身高：160 cm　　　　體重：50 kgs　　　健康：佳

教育背景：
2011-2013　　國立清華大學語言所碩士
2009-2011　　國立政治大學英文系
2005-2009　　文藻外國語文專科學校　主修英語／副修西班牙語

相關經驗：
2011-2013　　兼差翻譯
2009-2011　　英文家教
2007-2008　　英文系系學會會長
2006-2007　　學生代表會主席

Language ability：英文：　聽、說、讀、寫均優異
　　　　　　　　　　　中文：　聽、說、讀、寫均優異
　　　　　　　　　　　西文：　聽、說、讀、寫均良好

電腦技巧：Win 7, Word, Excel, Lotus

興趣：電影，音樂

希望待遇：NT$30,000 ／月

Date Available：7 月 1 日

〜〜

❷·❷

自傳

　　應徵工作時，除了履歷表外，自傳也是非常重要的書面資料。自傳的主要功能在對自己做一概括性的介紹。畢竟主考官從來沒見過應徵者，更無從認識應徵者。主考官可透過自傳對應徵者有初步的認識，而從眾多的自傳中挑出適合者參加面試，並在面試時，針對自傳、履歷表或其他資料提出相關的問題。由此可知自傳的重要性。

　　自傳通常像一篇短文，也可分成三至四段來寫，即所謂的起承轉合。內容大致包括了家庭背景介紹、個 描述、學經歷描述、特殊專長描述、請求給予面試的機會等等。換言之，自傳就像包裝紙，一篇令人印象深刻的自傳，在應徵工作時，扮演了催化劑的角色。

自傳㈠：有工作經驗

AUTOBIOGRAPHY

My name is Richard Lee, a married man who is 35 years old. There are 6 people in my family: my parents, my wife, two children and me. My parents are working as bankers and my wife is an art teacher. My major in Cheng Chi University was International Trade.

As I am very interested in global trade and I had very **good performance** while I was an undergraduate, I decided to **go further study**. I learned a lot about how to manage an organization while I was studying for the Master degree of Business Administration in Taiwan University. The knowledge that I obtained in the two universities did **help me a lot in** my further career.

My first job was working as a Special Assistant to G.M. in P.E. Co.. This job helped to practice all what I had learned in School. **Working as** a Special Assistant to G.M., it provided me a lot of opportunities to **expose to** international affairs. Thus, I decide to accept more challenges in my career by working as a Sales Manager in M. E. Co.. At this position, I need to negotiate with customers **all over the world**. I also traveled a lot to visit them and to attend international exhibitions. The sales **grow double** within three years when I ruled the Sales Department in M.E. Co.. As a friend of mine planned to expand his business, I was invited to be the G.M. in his branch office. Working as a G.M. in G.E. Inc., I was able to govern the branch office **from red to profit** within 4 years. All this performance **contributed to** the study and **practical experiences in the past**. In addition to that, my diligence and the assistance from my team member are also **the key to** the above mentioned success.

As the society and the global trend is **shifting rapidly and dramatically**, I am also **taking** some business **courses** at night to **be able to cope with** the changes. I believe that with my knowledge, experiences in the past and a heart to self break through, I can **be of great contribution to** your company.

❖ 譯文

自傳

我名叫李理查，現年35歲，已婚。我家共有六人：父母親、內人、兩個小孩和我自己。雙親皆任職於銀行，內人則是美術老師。我在政大主修國際貿易，因為我對國際貿易非常感興趣，所以在校成績表現非常優異，也因此決定畢業後再繼續攻讀碩士學位。在臺大攻讀碩士學位的過程中，學到了許多如何管理組織的相關專業知識。我在這兩間大學所習得的知識，對我日後職場生涯確實有相當大的助益。

我的第一份工作是總經理特別助理。這個工作提供了我將學校所學的各種知識實際運用的機會。藉由這樣的工作，我也得以有很多機會使自己接觸到複雜、有趣的國際事務。這也是我之所以決定接受一項新挑戰——接任業務經理的主要原因。這個職位需要與世界各地的客戶做各種協商，所以我常旅行於世界各地，一方面是拜訪這些客戶；另一方面則是參加不同的國際展覽。在我擔任業務經理期間，公司的業績在三年內成長了兩倍。後因朋友擴張事業版圖，故邀我轉任他分公司的總經理。我在擔任總經理一職的四年內，也將公司的營業由虧轉盈。這些成就除了歸功於先前所學的專業知識和實務經驗外，也與我自身的努力和經營團隊的通力合作有關。

有鑒於社會和世界趨勢瞬息萬變，我也利用晚上修習一些課程，以應付整個大環境的改變。我深信以我的專業知識、過去的經驗和一顆不斷自我突破的心，必對貴公司有所貢獻。

相關詞彙說明

good performance　良好的表現；良好的績效

go further study　進修；延伸學習

help somebody a lot in　在……方面對某人幫助很多

work as　擔任……的職務

expose to　暴露於……的環境中（指有接觸某事物的機會）

all over the world　世界各地

grow double　倍數成長

from red to profit　由虧轉盈（red 是赤字的意思；profit 是利
　潤的意思）

contribute to　歸因於；歸功於（也可用 attribute to 來表達）

practical experience　實務經驗

in the past　在過去

the key to　……的關鍵因素

shift rapidly and dramatically　劇烈變化

take .... courses　上……的課程

be able to　能；可以（也可用 can 來取代）

cope with　適應

of great contribution to　對……有貢獻（也可用 contribute to
　來表達）

自傳㈡：社會新鮮人

AUTOBIOGRAPHY

My name is Jessica Fang. I was born in Kaohsiung in 1985. There are 7 people in my family. I have two brothers and two sisters. I am the 4th child in my family. My parents **regard** honesty and diligence **as the virtues** that shall be followed. Hence, I have been taught to be honest and diligent since I was a child.

As I **have talent in** language, I decided to study in Wen Tzao Foreign Language Junior College. There, I took English as my major and Spanish as my minor. In addition to **keeping good performance in** study, I also attended a lot of **extra curriculum activities** in W.T.. I **was** even **elected as** the head of both English Association and Student Union. Learning different languages **leads** me **to view** the world **globally** and to be able to appreciate different cultures. My experiences in extra riculum activities made me understand how to work with a group of people, how to organize a team and achieve our goal as we planned. This experience let me **see the importance of** interpersonal skills.

As I have great interest in languages, I decide to go further study in Cheng Chi University. I also decided to get a master degree in English Linguistics. To be able to **support myself**, I **took part time job** as an English tutor and translator. Being a tutor helped me to be able to express myself in a clear and logical way. As a translator, I was able to obtain knowledge **from different angles**.

I saw from the job website that you are looking for an global marketing specialist. As for my **career plan**, I hope my job will be more **internationally-oriented**. With my knowledge in two foreign languages, my interpersonal skills and personalities, I believe that I can be of great contribution to your company.

❖ **譯文**

〜〜〜〜〜〜〜〜〜〜〜〜〜〜〜〜〜〜〜〜〜〜〜〜〜

自傳

我名叫 Jessica Fang，於1985年出生於高雄。我家共有七個人。我有兩個姊妹和兩個兄弟，我在家排行第四。父母親視誠實與勤奮為必須遵守的美德，所以我自小就養成誠實和勤奮的態度。

因為我有語言天分，所以決定進入文藻外語學院攻讀語言。在那裡我主修英文，副修西班牙文。除了努力保持好成績外，我也參加了許多課外活動，還被選為英文學會會長和學生會主席。學習不同的語言讓我能以國際觀的角度來看這個世界，且也能從欣賞的角度來看待世界上不同的文化。我在課外活的歷練也讓我了解如何與一群人合作、如何組織一個團體並達成計畫中的目標。這些經驗都有助於我的人際關係技巧。

由於我對語言極有興趣，所以決定插班政治大學繼續進修，大學畢業後更決定繼續攻讀英國語言碩士學位。為了能自給自足，我兼職英文家教和翻譯的工作。英文家教的工作讓我以清楚和邏輯的方式來表達自己；翻譯的工作則使我可以從不同的角度學得知識。

我在工作網站上得知貴公司在招募國際行銷專員。我的生涯計畫是希望未來的工作能具國際導向。以我具備兩種外語能力的背景、人際關係的技巧和個性，我深信我可以對貴公司有所貢獻。

〜〜〜〜〜〜〜〜〜〜〜〜〜〜〜〜〜〜〜〜〜〜〜〜〜

相關詞彙說明

regard ... as virtues　將……視為美德

have talent in　在……有天分

keep good performance in　保持良好的表現、成績、績效

extra curriculum activities　課外活動

be elected as　被推選為

lead to　導向

view... globally　以國際觀的眼光來看……

see the importance of　重視……（也可用 see the importance of/regard/value... 來表達）

interpersonal skills　人際關係技巧

support oneself　自給自足

take part time job　兼差

from different angles　從不同的角度

career plan　生涯計畫；生涯規劃

internationally-oriented　國際導向

2·3

求職信

　　應徵工作除了必須準備履歷表、自傳外，還需準備一份求職信。求職信中的內容也大致分為三段：第一段說明從何處得知應徵的消息；第二段是本文，再次說明自己的確符合資格擔任此工作；第三段則是致謝並請求給予面試機會。

　　求職信的寫法，必須依照英文書信的格式來書寫，以下提供兩種範本供參考。

求職信㈠：有工作經驗

Richard Lee

No. 666 Ming Shen East Road, Taipei, Taiwan, R.O. C.

Tel: (02) 2772-2838　Mobil: 0911-111-111

E-mail: Richard@ms31.hinet.net

Date: March 7, 2013

ABC Electronics Inc.,

8F, Song Chiang Rd., Taipei, Taiwan R.O.C.

Tel: (02) 2366-8878

Att: Manager of Human Resources Dept.

Dear Sir,

I learned from the Job Search Websites about your esteemed Inc. is now **having an opening position for** Chief Operating Officer. I would like to take this opportunity to introduce myself to you as one of the candidates.

I have 9 years of working experiences **in this field**. It means that I have profound knowledge in electronics industry. I have been working as a Special Assistant to G.M., then get promoted to be a Sales Manager, and then to a Director. With these past experiences, I was not only be able to run a team but also bring benefit to a company within years. **With references to** my resume and autobiography, you will have **a clear profile** about my personalities and competence.

Should you find my **qualifications meet with your requirements**, I shall **appreciate** a lot if you would **grant me an opportunity for** an interview. I **am confident to** prove my knowledge and ability to contribute to your **organization**.

Thank you for taking your time. Looking forward to your kind reply.

Yours faithfully,

Richard Lee
Richard Lee

Enc.: 2

❖ **譯文**

〰〰〰〰〰〰〰〰〰〰〰〰〰〰〰〰〰〰〰〰〰〰〰〰〰〰〰

　　我從求職網站中得知貴公司營運長的職位出缺。我想藉此機會向貴司推薦自己為合適的候選人之一。

　　我在這個領域有九年的工作經驗，這表示我在電機工業有相當深厚與淵博的專業知識。我曾擔任過總經理特別助理的工作，而後被擢昇為業務經理和處長。藉著這些過去的工作經驗，我不僅有機會得以帶領團隊，而且多年來也為公司帶來了獲利。貴公司可從我的履歷表和自傳中得知有關我的能力和個　的相關資料。

　　如果貴公司認為我的資格符合貴公司的要求，敝人將對貴公司惠賜的面試機會感激萬分。我有自信可以以一己的專業知識和能力對貴公司有所貢獻。

　　謝謝貴公司花時間閱讀這份資料！期待收到貴公司的善意回覆！

〰〰〰〰〰〰〰〰〰〰〰〰〰〰〰〰〰〰〰〰〰〰〰〰〰〰〰

相關詞彙說明

have an opening position for　　有……的職缺

in this field　　在此領域中

with reference to　　與……相關

a clear profile　　清楚的輪廓／概念

qualification　　資格

meet with your requirements　　符合您的要求

appreciate 感激；欣賞

grant sb. an opportunity for 惠賜某人……的機會

be confident to ＋原型 詞 有自信可以做……

organization 組織；機構；公司（在商業書信中，通常泛指
公司而言）

求職信㈡：社會新鮮人

Jessica Fang

No. 666, Shan Ming Rd., Shan Ming District, Kaohsiung, Taiwan,
R.O. C.

Tel: (07) 7772-2838 e-mail: **jessicaf@ms54.hinet.net**

Date: July 9, 2013

Kimberly Global Marketing Co., Ltd.

No. 115, Chung Shiao East Rd.,

Taipei, Taiwan, R.O.C

Att: Mr. Howard Huang/Manager of Personnel Dept.

Dear Mr. Huang,

It is my pleasure to learn that **there is an opening for** Global

Marketing Specialist in your **esteemed company** through 104 Job Website. **I am submitting my application for your consideration.**

My talent and knowledge in foreign languages is one of the key to be a global marketing specialist. My educational background **enables** me to obtain a global **view**, and to be able to **adapt to** different cultures and **thinking patterns**. My experiences in curricular activities also help to **sharpen** my interpersonal skills. My experiences as a translator made me able to see things from different angles.

For your better evaluation, please see the enclosed resume and autobiography of mine. I shall appreciate you giving me a chance for an interview when I am confident to **prove** my knowledge and ability to contribute to your esteemed company.

Thank you for taking your time reading my **data**. **I look forward to** hearing your **warm reply**.

Sincerely yours,

Jessica Fang
Jessica Feng

Encl.: 2

❖ **譯文**

〰〰〰〰〰〰〰〰〰〰〰〰〰〰〰〰〰〰〰〰〰〰〰〰〰〰〰〰〰〰

　　很榮幸從 104 工作網站上得知貴公司目前國際行銷專員的職位有空缺。在此提供我個人的申請資料以作為貴公司的參考。

　　我個人在外語的天分和專業知識是成為國際行銷專員的關鍵之一。我的教育背景使我具有國際觀，且使我易於適應不同的文化和思考模式。敝人在課外活動上的經驗也增進我在人際關係上的技巧；而翻譯的經驗讓我能從不同的角度來看事情。

　　附件的履歷資料和自傳將有助於貴公司做進一步的評估。若貴司能惠賜面試的機會，我將感激不盡。我也有自信證明我的專業知識和能力能對貴公司有貢獻。

　　謝謝貴公司抽空閱讀我的資料，期待收到貴公司之善意回應

〰〰〰〰〰〰〰〰〰〰〰〰〰〰〰〰〰〰〰〰〰〰〰〰〰〰〰〰〰〰

相關詞彙說明

It's my pleasure to+ 原型動詞　很榮幸……，也可用 It's a
　　pleasure to + 原型動詞來表達

esteemed company　貴公司（前面加 esteemed 來表示尊稱和
　　尊敬，是一種客氣用法）

there is an opening for　有空　（也可用 have an opening
　　position for 來表達）

I am submitting my application for your consideration.
　　在此提供個人的申請資料供貴公司參考（也可用 "I
　　would like to take this opportunity to present/introduce

myself as a candidate." 來表達）

enable　使能；使可以

view　n. 觀點 v. 看

adapt to　使適應（也可用 adopt（接受）來取代。）

thinking patterns　思考模式

sharpen　加強；改善（原意是使尖銳、使鋒利）。

data　資料（也可用 information 來取代）

look forward to　期待（look forward to + Ving，這個片語後
　　面只能加動名詞）

warm reply　善意的回應（也可用 kind reply 來取代）

2·4

回覆求職信函

　　公司在審核完應徵者的相關資料或在面試完後，會發布錄取
或未錄取通知。以下列舉不同案例，供讀者參考。

通知面試

LETTERHEAD

July 9th, 2013

Ms. Amanda Chen
3F, No. 222, Chung Cheng Rd.,
Chong Ho City, Taipei, Taiwan

Dear Ms. Chen,

Sub: **Notification of Interview**

Thank you for replying to our **advertisement** in 104 Job website in June for the opening position of Marketing Director. We are pleased to **keep you informed that** your resume has been received and **reviewed** carefully.

We found that your qualification **perfectly** meet with our requirements. We would like to invite you for **an interview** at 10:00 AM. of July 14. We look forward to seeing you then. Please **be on time** and **bring** your work **with you**. Wishing you good luck.

Yours sincerely,

Emily Dickson
Emily Dickenson
General Manager

❖ 譯文

〰〰〰〰〰〰〰〰〰〰〰〰〰〰〰〰〰〰〰〰〰〰〰〰

主旨：面試通知

謝謝您回覆我們在 104 工作網站上所登的徵求行銷經理的廣告！很榮幸通知您我們已收到您所寄來的履歷資料，且我們也已仔細的閱讀過資料。

我們發現您的資格相當符合我們的要求，想請您在七月十四日上午十點到敝公司面試。期待您的到來，並請您隨身攜帶您的作品。

祝您好運！

〰〰〰〰〰〰〰〰〰〰〰〰〰〰〰〰〰〰〰〰〰〰〰〰

相關詞彙說明

notification of interview　面試通知（也可用 advice of interview 來表達）

advertisement　廣告（可用縮寫 advert. 或 Advt. 來取代）

keep sb. informed that + 子句　通知某人……（也可用 keep sb. advised that ...）

review　(v.) 審核；審閱；討論

perfectly　非常；相當（原意為完美地、十全十美地）

an interview　面試

be on time　準時（也可用 be punctual 來取代）

bring with sb.　隨身攜帶

work　(n.) 作品；(v.) 工作

通知錄取㈠

LETTERHEAD

Date: August 14, 2013

Mr. Edmund Kimberly

P.O. Box 94,

Taipei,

Dear Mr. Kimberly,

Through the **documentary review, written test** and interview, you have **made a very good impression on** us.

With **a cheerful heart**, we would like to inform you that our **board members** have **finally decide**d to **have you for the position of** Branch Manager of **branch office** in Taipei. We believe that your **leadership** will **benefit our branch office a lot**.

We would like you to start working from the first day of next month. We look forward to seeing you then. Should you have any **queries**, please feel free to contact me **at your convenience**.

Truly yours,

Catherine Binoche

Catherine Binoche

Head of Group Development

❖ 譯文：

在文件審查、筆試、面試等過程，我們對您的表現留下了深刻的印象。

在此很高興的通知您，敝公司董事會成員最後決定由您擔任我們臺北分行行長一職。我們深信您的領導統馭將對我們分公司有很大的助益。

我們希望您從下個月一號開始工作，期待到時能見到您。如果您有任何的疑問，請您於方便的時候與我連絡。

相關詞彙說明

through　透過；藉由（也可用 by means of/by）

documentary review　文件審查

written test　筆試（口試則是 oral test）

A made a very good impression on B　（B 對 A 留下深刻的印象）

a cheerful heart　一顆興高采烈的心

would like to + 原型動詞　想要

inform　通知；告知（也可用 advise/tell/let you know 來替代）

board member　董事會成員

finally decide to + 原型動詞　最後決定

have sb. for the position of　由、讓某人擔任……的職位

leadership　領導統馭

benefit ... a lot　對……助益良多

query　n./v. 詢問；疑問；質問

at your convenience　於您方便的時候（是一種英文書信常見的禮貌用語）

head of ...　……的負責人

通知錄取㈡

Letterhead

Date: Mar. 10, 2013

Mr. Robin Hu,

4/F, 19 Canton Fair Rd.,

Tsim Sha Tsui,

Wan Tzai, Hong Kong

Dear Mr. Hu,

Thank you very much for the time you spent in **applying for** employment with our company.

All the **interviewers spoke very highly of** your performances during the conversation. We have **come to a consentient conclusion** that you are **the right person for** the opening position of head of R&D Group.

The **terms** and **benefits** are as we discussed **during the interview**. We shall **expect to** see you start working in two weeks. If you **decide not to** come, please **call us at** (07) 2332-2868

Sincerely yours,

Jessica Fang

Jessica Fang

Personnel Manager

❖ 譯文：

〰〰〰〰〰〰〰〰〰〰〰〰〰〰〰〰〰〰〰〰〰〰〰〰〰〰〰〰

　　非常謝謝您抽空參加敝公司的應徵活動！

　　所有的面試主考官在談話中都對您有相當高的評價，並一致推崇您是目前出缺的研發部負責人的最佳人選。

　　工作條件和福利都如我們面試時所談的內容。我們期待在兩週後看到您開始擔任這個工作。如果您決定不來敝公司上班，請打 (07) 2332-2868 與我們聯絡。

〰〰〰〰〰〰〰〰〰〰〰〰〰〰〰〰〰〰〰〰〰〰〰〰〰〰〰〰

相關詞彙說明

apply for　申請……；應徵……

interviewer　面試者；面試主考官

speak very highly of sb.　對某人有相當高的評價

come to a conclusion　達成結論

consentient conclusion　一致的結論、共識

the right person for　……的恰當人選

terms　工作條件（可用 working terms/conditions 來替代）

benefits　福利（也可用 welfare 來表達）

expect to+ 原型動詞　期待……（to 後面只能加原型動詞）

in two weeks　在兩週後（in + 時間片語，表示未來的時間）

decide not to + 原型動詞　決定不……（to 後面只能加原型動詞）

call us at + 電話號碼　打電話聯絡我們（電話號碼前要加介詞 at）

不錄取通知㈠

Letterhead

Date: Mar. 10, 2013

Ms. Ann Lee
3/F, 25 Brower Ave.
Kowloon, Hong Kong

Dear Ms. Hu,

Thank you very much for **taking the trouble to** call and to send us your resume.

Your **excellent background** and **expertise** are **highly acclaimed**. However, many other highly **qualified people** also applied for the positions, which made the choice extremely difficult. Therefore, we **regret to** inform you that you were not the one selected this time.

We will keep your application on active file for future opening. **In the meantime, we wish you success in your future endeavors.**

Sincerely yours,

Dona Lin

Dona Lin

Manager

❖ 譯文：

謝謝您特意打電話並寄上您的履歷資料。

您優異的背景和專業頗受敝公司肯定。然而，同時間也有許多相當符合資格的人應徵這份工作，讓甄選變得相當困難。所以我們很遺憾通知您這次並未被錄取。

我們會將您的應徵函保留在「有效檔案」中，以作為將來有職缺時的優先選擇。同時，祝您在未來的應徵一切順利。

相關詞彙說明

take the trouble to + 原型動詞　特意地（to 後面只能加原型動詞）

excellent background　優異的背景

expertise　專業知識；專業技能

highly acclaimed　受到高度讚揚、肯定

qualified people　符合資格的人選

regret to + 原型動詞　遺憾（to 後面只能加原型動詞）

in the meantime　同時（也可用 at the same time; meanwhile 等用辭）

endeavors　(n.) 努力，盡力。(v.) endeavor to + 原型動詞

We wish you success in your future endeavors.　祝您在未來的應徵一切順利。（這是一個相當常見的制式化表達用語，專用於祝福他人應徵工作順利成功。）

不錄取通知㈡

Letterhead

Date: Mar. 11, 2013

Ms. Daisy Huang
No. 443 5th Ave.
Manhattan, NY.
U.S.A.

Dear Ms. Huang,

Thank you very much for your application for the position of Sales Assistant at the Marketing Dept.. Unfortunately, we regret to advise that your application was not successful **on this occasion**.

Please understand that this decision is **in no way** an **unfavorable reflection on** your **accomplishment**. We have received 100 applications for only one **vacancy**. **Inevitably**, we must **disappoint** many **well-qualified** candidates.

Thanks for your time and **effort**. **Wishing you the best for the future endeavors**.

Sincerely yours,

Dona Lin

Dona Lin

Manager

❖ **譯文：**

〰〰〰〰〰〰〰〰〰〰〰〰〰〰〰〰〰〰〰〰〰〰

　　非常謝謝您應徵行銷業務助理一職。很遺憾通知您這次的應徵未果。

　　希望您瞭解這個決定絕非表示否定您過去的成就，只是我們只有一個職缺，卻收到一百份應徵函。因此讓許多符合資格的候選人失望是無法避免的。

　　非常謝謝您花費的時間和努力。祝您在未來的應徵一切順心如意。

〰〰〰〰〰〰〰〰〰〰〰〰〰〰〰〰〰〰〰〰〰〰

相關詞彙說明

on this occasion　在這個時機；機會

in no way　決非；決不

unfavorable　不利的；反對的

reflection on　對……的反應（也可用 feedback 來取代 reflection）

accomplishment　成就（也可用 success 或 achievement 來取代）

vacancy　出缺；空位（也可用 opening/available position 來取代）

inevitably　不可避免地

disappoint　使失望（這裡也可用 refuse/reject/to say no/turn down/decline 等詞來取代，以上的字意都是拒絕的意

思。只是用 disappoint 較客氣且較不直接，比較不傷人）

well-qualified 符合資格（亦可用 highly qualified 來取代）

effort （n.）努力／盡力（動詞片語為 make effort 或 do the best）

Wishing you the best for the future endeavors. 祝您在未來的應徵一切順利。（這是一個相當常見的制式化用語，專用於祝福他人應徵工作順利成功。也可用前面所使用的 "We wish you success in your future endeavors." 來表達同樣的意思。）

開發信

商業行銷有很多方法，開發信（又稱為推銷信）是運用最廣的方式之一。開發信的重點在於向潛在客戶介紹公司的特色或產品，最終的目的是要引起潛在客戶的興趣、刺激客戶的需求、進而採取行動。

3·1 AIDA 原則

推銷信是非常具挑戰性的一種信函，因為書寫者的重點是希望透過這封信，來引君入甕，使收信者依據自己的需求採取後續的相關行動。所謂 AIDA 是以下四個英文字的縮寫：

1. **Attention**（**引起對方的注意**）：是指透過精心設計的標題或畫面來吸引對方的注意。這是第一步也是最重要的一步，如果無法吸引讀者的注意，也就不會有後續的動作發生。

2. **Interest**（**激發對方的興趣**）：是指透過精心設計的標題、畫面、內容來引起對方的興趣。

3. **Desire**（刺激對方的需求）：是指文章的內容除了要能引起對方的注意、激發對方的興趣外，接下來就是要刺激對方的需求，讓對方覺得他的確有此需要。

4. **Action**（採取行動）：刺激對方的需求還不夠，一封具說服力的信，最終目的就是要讓對方依寫信者的期望採取行動。

3·2 撰寫與回覆開發信

收信者在收到開發信後，若感興趣就可針對推銷信的內容進一步的詢問；若一點也不感興趣，則不用作任何回覆。當然你也可以回覆不感興趣的推銷函。

實例一 推銷公司之服務與相關設施

Letterhead

Date: Mar. 11, 2013

Franze Hegel Co., Ltd.Victory

Leimenstrasse 28

CH-2008 Basel

Switzerland

Sub.: Don't Miss The Chance! Come and Stay With Us!

We got your esteemed company name from the hotel exhibition in Dusseldorf. We would like to **take this opportunity to** introduce you our hotel, Golden Empire Hotel.

We are a hotel of 200 years history. We **are very famous for** our **high quality** service and **fully-equipped** facilities. We have 800 quest rooms, 20 small **conference halls**, 15 medium conference halls and 10 large conference halls, which can **accept** 8 - 100 persons. We also have two gym., two swimming pools, two SPA and two Sauna which open from 10:00AM ~ 24:00PM. **In addition**, we **have** several restaurants **for your choice**. We also have a 24-hour **business center** where you can send e-mail, fax and use internet. We have **staff** who can speak various languages. Language **is never** a **barrier** in our hotel.

Next time when you come to Hawaii, don't forget to come and stay with us. We believe that you will **have wonderful time** being with us, and you would like to come back again.

Make the reservation now! If you need any information, please contact us or **visit our website**.

Sincerely yours,

Teresa Chen
Teresa Chen/Marketing Director

❖ **譯文**

〰〰〰〰〰〰〰〰〰〰〰〰〰〰〰〰〰〰〰〰〰〰〰

主旨：別錯過機會！請來與我們同住！

　　我們從杜塞道夫的飯店展覽中獲得貴公司的相關資料。藉此機會，我們想向您介紹我們的飯店——黃金帝國大飯店。

　　黃金帝國大飯店擁有兩百年的歷史，以高品質的服務和設備齊全而聞名。我們有八百間客房、二十間小會議廳、十五間中型會議廳，這些會議廳可容納 8 -100人。我們也有兩間健身房、兩間商務中心、兩個游泳池、兩間水療室和兩間桑拿室，開放的時間是早上十點到深夜十二點。此外，我們也有許多餐廳供您選擇。我們更有24小時開放的商務中心供您發送電子郵件、傳真和上網。還有會說多國語言的工作人員，所以語言在我們的飯店從來不是障礙。

　　下次當您來夏威夷時，別忘了造訪我們飯店。我們相信您會與我們共度美好時光，而且保證您還會想再回來。趁現在趕快預約住房吧！如果您需要任何資料，請與我們聯繫或造訪我們的網站。

〰〰〰〰〰〰〰〰〰〰〰〰〰〰〰〰〰〰〰〰〰〰〰

相關詞彙說明

Don't miss the chance. 別錯失機會

take this opportunity to + 原型動詞 趁此機會……（to 後面加原型動詞）

be famous for 以……聞名／著名（也可用 be well-known/widely known for 來取代）

high quality 高品質

fully-equipped 設備齊全

conference halls 會議廳（也可用 conference room 來取代）

accept 原指接受，這裡引申為容納。

in addition 除此之外

have ... for your choice 有……供選擇

business center 商務中心

staff 職員；工作人員

be never 從不、絕不……（也可用 be in no way 來取代）

a barrier 障礙（也可用 an obstacle 來取代）

have wonderful time 玩得盡興（也可用 have a good time 或 have fun）

visit our website 造訪我們的網站

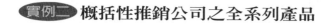

概括性推銷公司之全系列產品

Letterhead

Date: Mar. 10, 2013

I. V. Co.,Ltd.
27 Clair Avenue East
Toronto, Ontario
M5S 3M2 Canada

Att: Importing Manager

Dear Sir,

We learn from Canada Trade Center that you are **looking for** industrial valve supplier from Taiwan. We would like to take this opportunity to introduce you about our company.

Global Valve Co., is a **leading** thermoplastic valve **manufacturer** in Taiwan. We **specialize in** industrial valves, such as butterfly valve, swing check valve, diaphragm valve and Double Union ball valve. Our products **meet different standards**, such as ASTM, BS, DIN, JIS and CNS standards. We are an ISO, LPCB, FM, UL and NSF

certified company. We have three manufacturing plants in different countries: Taiwan, China and Malaysia. We have 500 employees in total. Among the total employees, we have 30% of **human resources in R&D**. This means that we are **a company of innovation**. We own 50 **patents** all over the world. We export our products all over the world. We **have high reputation in** quality and services.

We would like to **express our interest to** be the supplier that you are looking for. We **have full confidence** that we will meet all your requirements. Enclosed is our **full set of** catalogues. We look forward to entering into **business relationship** with you.

Sincerely yours,

Andrew Wei

Andrew Wei/General Manager

Encl.: full set catalogues

❖ 譯文

〜〜〜〜〜〜〜〜〜〜〜〜〜〜〜〜〜〜〜〜〜〜〜〜〜〜〜〜〜〜〜

　　我們從加拿大世貿中心得知貴公司正在找尋工業閥的臺灣供應商。我們想藉此機會向您介紹有關敝公司的相關訊息。

　　國際閥公司在臺灣是一家具領導地位的塑膠閥製造商。我們專精於工業閥的生產，如蝶閥、擺動型逆止閥、膈膜閥和雙由令球閥。我們的產品符合美規、英規、德規、日規和中華民國規格等不同標準，同時是一家ISO、LPCB、FM、UL和NSF認證合格的公司。我們在不同的國家共有三個製造工廠：臺灣、中國大陸、馬來西亞，員工共五百名。在所有的員工中，30％為研發人員，這表示我們是一家具創新能力的公司。我們公司在世界各地擁有五十種以上的專利權，產品輸出遍及全世界，不論在品質或服務都頗有口碑。

　　我們對成為貴公司正在找尋的供應商具有興趣。我們相當自信可以完全滿足貴公司的要求。附上敝公司的完整目錄供貴公司參考，並期待與貴公司建立商業關係。

〜〜〜〜〜〜〜〜〜〜〜〜〜〜〜〜〜〜〜〜〜〜〜〜〜〜〜〜〜〜〜

相關詞彙說明

look for　找尋

leading manufacturer　具領導地位的製造商

specialize in + Ving　專攻於；專精於（也可用 be good at + 名詞的說法來取代）

meet different standards　符合不同的標準

certified company　認證合格的公司

human resources　人力資源

a company of innovation　一家具創新能力的公司

patents　專利

have high reputation in　在……方面有很好的聲譽（可用 be widely known for）

express our interest to + 原型動詞　表達……的興趣（也可用 express our interest in + 名詞或動名詞）

have full confidence　有十足的信心（也可用 have complete confidence）

full set of　完整的；整套的

enter into relations　建立關係（也可用 establish relations）

business relationship　商業關係

實例三 推銷公司特定產品

Letterhead

Date: Mar. 15, 2013

Bike Incorporation
1100 Carlton Boulevard
Topeka, Kansas, 33443
U.S.A.

Att: Mr. Waller V. Walker/Purchaser Manager

Dear Waller,

Sub: The Latest Model of Bike

Over the past 10 years, you have been giving your support and cooperation. We value such business relations with you, and would like to thank you for all.

We have **come up with** a latest model this month: Model 1238, **auto-controlled all-terrain bicycle**. This bicycle **is especially made for** the **professional mountaineer**. It is light and small, could be folded and put into a backpack. It can perfectly function in all kinds of terrains, such as **downhill**, **uphill**, **winding roads**, **bumpy paths** and **steeps**. Once you **mount** it, everything will be auto-controlled.

We believe that you would **be more than satisfactory with** this

newly invented model. Please find the **attached information** and our best offer for this mode. You can also visit our website for further information.

We look forward to hearing your kind reply.

Sincerely yours,

Edith Lee
Edith Lee/Head of R&D Center

❖ 譯文

主旨：最新款式的自行車

　　過去十幾年，您一直給予敝公司支持和合作。我們很珍惜且重視與您之間的商業關係，並對所有的一切向您致意。

　　我們在這個月推出了一種新的款式：1238 型──適合各種地勢的全自動自行車。這款自行車是專為專業的登山者所研製，不僅輕巧且可摺疊放入背包，不論在各種地勢如：上坡、下坡、九彎十八拐的道路、崎嶇不平的羊腸小徑和陡峭的坡道等都可運作自如，只要騎上這部自行車，一切將會自動控制。

　　我們相信您必定對此新研發的款式感到非常滿意。請參考附件的資料和我們最優惠的價格。您也可以造訪我們的網站以取得更進一步的訊息。

　　期待您的善意回應。

相關詞彙說明

the latest　最新的；最近的

over the past 10 years　過去十幾年（也可用 for the past 10 years）

come up with　原指趕上、提供、準備之意，這裡引申為提供之意。

auto-controlled　全自動

all-terrain　各種地勢、地形

be especially made for　特別為……而製（被動式的表達）

professional mountaineer　專業的登山者

downhill　上坡

uphill　下坡

winding roads　winding 是彎曲、曲折之意，故譯成九彎十八拐，使讀者更易瞭解。

bumpy paths　bumpy 是崎嶇不平、顛簸之意；path 是小道之意，譯成羊腸小徑更是入木三分。

steep　陡峭的坡道

mount　騎上（也可用 ride 這個動詞）

be more than satisfactory with　感到非常滿意

newly invented　新研發的

attached information　附件資料，通常傳真時的附件用 attached 來表達；寄信時的內附文件用 enclosed 來表達，也可用縮寫 encl. 來取代。兩者不可混淆。

實例四 表示對推銷信感興趣

Letterhead

Date: Mar. 15, 2013

Bike Incorporation.

1100 Carlton Boulevard,

Topeka, Kansas, 33443

U.S.A.

Att: Mr. Waller V. Walker/Purchaser Manager

Dear Flowrecnce.

Sub: The Latest Model of Bike: Model 1238

Let me begin by thanking you for keeping me informed of your latest Model. I **acknowledge** that I also received your attached information. I also want to confirm that I am **definitely** interested in this new model.

Pursuant to the conversation we had this morning, please send me some photos of this model. **It would be best if you could** send

me the pricing and one sample **as soon as possible**. This would **permit** us to **evaluate** the **marketability**. **As thing stands now**, we are interested in **pursuing** the market share of this new model as **there is no** other **competitor** in the market.

After our **marketing research**, if it proves to be positive, I would fly to you for **further discussion**. We look forward to hearing your kind reply.

Sincerely yours,

Miranda Chen

Miranda Chen/Purchasing Mgr.

❖ **譯文**

　　首先感謝您通知我有關貴公司的最新款產品。我確實已收到了您所附上的相關資訊，且對此新款產品相當感興趣。

　　如我們今天早上所談的，請寄給我這型號的相片。如果您可以盡快將報價和樣品寄給我是再好也不過，這樣可以使我們評估其市場可行性。就現階段而言，我們看好這項新產品的市場佔有率，因為到目前為止，市場中沒有其他的競爭者。

　　在市場調查後，如果獲得正面回應的話，我將會飛往您那裡與您做進一步的討論。期待您的善意回覆。

相關詞彙說明

let me begin by thanking you for　首先為……向您致謝（這是一個很實用的開頭語）

acknowledge　告知已收到（某物）

definitely　明確地；確切地；一定；肯定（用來加強語氣）

pursuant to　依據；依照（可用 according to/base on 來替代）

it would be best if you could...　如果你可以……將再好也不過（一種請求對方的客氣委婉說辭）

as soon as possible　盡快（可用 immediately/promptly/urgently 來取代）

permit　促使

evaluate　評估

marketability　市場可行性；銷路

as thing stands now　就現階段而言（可用 at this very moment 來取代）

pursue　追求；從事；進行

market share　市場佔有率

there is no + 名詞　沒有……（後面只能加名詞）

competitor　競爭者

market research　市場調查（也可用 market investigation）

further discussion　進一步的討論（可用 more discussion 來替代）

實例五 對推銷信感興趣，並索取更多的訊息

Letterhead

Date: Mar. 16, 2013

Golden Empire Hotel
1000 Freemental Ave.
Honolulu, A23553
Hawaii

Sub.: Need More Information of Your Esteemed Hotel

Dear Sirs,

I received a promotional letter **along with** some other **useful information** from you. I read all the **materials provided by** you and **find out** that we could apply for a membership from you.

We are a global **investment company**. We have **thousands of investors** all over the world. We need to **hold** many **conferences** and **seminars to** all investors. As Hawaii is famous for the pleasant weather and beautiful landscape, we decide to hold the important **event** there annually. You mentioned in your letter that there is no language barrier in your hotel, and **that's why** we are very interested in **working with** you **in the very near future**. We need more information from you, such as complete brochure, **membership program**, airport-picking-up program, **special discount** program**... and so on.**

Your prompt reply will be highly appreciated. Looking forward to entering into a **substantial** business relationship with you.

Sincerely yours,

Joan Tseng
Joan Tseng
Director of P. R. Dept.

❖ 譯文

主旨：需要更多有關貴飯店的資訊

　　茲收到貴公司所寄來的推銷函和其他實用的相關資料。讀過所有資料後我發現我們可以向貴公司申請加入會員。

　　我們是一家跨國投資公司，在世界各地有數千名的投資人，且需要為這些投資人舉辦多場的會議和說明會。由於夏威夷以宜人的氣候和美麗的風景聞名於世，所以我們決定每年在那裡舉辦重要盛事。在您的信函中提到貴飯店沒有語言障礙，這正是我們屬意不久的將來能與貴飯店合作的主因。我們需要您提供更多的資訊，如完整的型錄、會員活動方案、機場接送服務、特殊折扣方案等。

　　若您能快速回覆將不勝感激。期待與貴飯店建立實質的合作關係。

相關詞彙說明

along with　與……一起；……以外（可用 and 或 with 來取代）

useful information　實用的資料（也可用 practical information）

materials　原指原料、材料，這裡泛稱所提供的各種資料。

provided by　由……所提供

find out　發現（這裡也可用 discover 來取代）

apply for a membership　申請會員

investment company　投資公司

thousands of　數以千計的

investor　投資者；投資大眾

hold conferences to/for 人　為了……舉辦會議

seminar　說明會；專題討論會

event　盛事；大事

that's why + 子句　那是為什麼；那是……的主因（後面加上說明的子句）

work with　與……合作（這裡也可用 cooperate 來取代）

in the very near future　在不久的將來

membership program　會員活動方案

special discount　特殊折扣

... and so on　等等

substantial　實質的；大量的；重要的

實例六 客氣的回覆不感興趣的推銷信

Letterhead

Date: Mar. 13, 2013

Global Valve Co., Ltd.

243 Chung Cheng Rd.

Kaohsiung , Taiwan

R.O.C.

Dear Sirs,

Thank you for your letter **dd.** March 10, in which you **show your interest in** being our supplier.

The letter that you send to our importing manager was forwarded to me. It is unfortunate that your information did not arrive **prior to** the moment we had made our **final decision on** picking a new supplier. **Had we been in receipt of your information before making decision, our response would have been different.**

Our staff has studied all the information provided by you. **There is no doubt that** your esteemed company is a qualified supplier **in our**

sphere of activity, **in terms of** product lines, various specifications, quality control, **ability to innovate as well as** high reputation you have achieved in the world so far. Unfortunately, we have already **signed a contract for a period of** two years with another qualified supplier in your region.

Nevertheless, your information is very much appreciated and will be kept on file **for future reference**. We wish you all possible success in your future endeavors.

Sincerely yours,

Cecilia Lee

Cecilia Lee

Purchasing Mgr.

❖ **譯文**

感謝您三月十號來函表示有興趣成為敝公司的供應商。

您寫給我們進口部經理的信已轉交給我。很不幸的,這封信並未在我們選擇供應商定案前抵達。如果我們定案前就收到您的來函,結果應該會不同。

我們的成員已研讀了您所提供的所有資料。無庸置疑,貴公司在產品線、多樣化的規格、品質控制、研發創新的能力,以及貴公司到目前為止在世界上所建立的知名度等,顯然是我們這領域中合乎資格的供應商。不幸的，我們已經與另外一位和您同一地區的供應商簽訂了兩年的合約。

　　但是，我們很感激您所提供的資料，並會將之存檔，以供未來參考。祝您未來一切順遂。

相關詞彙說明

dd.　其上所標註的日期為……（是 dated 的縮寫；也可用 of ＋日期來取代）

show your interest in　表示對……有興趣（可用 express interest in ... 取代）

prior to　先前；在……之前（可用 before/in advance 取代）

final decision on　在……做最後決定

in receipt of　介詞片語，收到的意思（可用 receive 這個動詞來表達）

make decision　做決定（可用 make selection/make choice 來取代）

response　回覆；回答；回應（也可用 reply/answer）

Had we been in receipt of your information before making deci-sion, our response would have been different.　這是一個倒裝句，也可換成 "If we had received your information ahead of our decision, our reply would be much different."

There is no doubt ＋ 子句　無庸置疑地（也可用 Undoubtedly/ Unquestionably） in our sphere of activity　在我們的領域中

in terms of　在……方面；從……方面來看（也可用 from the
　　aspects of ...）

ability to innovate　創新能力

as well as　和；以及（可用 and 取代）

sign a contract　簽訂合約

for a period of　為期……的

for future reference　供未來參考

Making Inquires

詢問

　　為了使商業行為進行順暢，通常買方會針對產品、樣品、價格、交期、運送方式、包裝、品質、付款條件等提出疑問，以便評估是否進行下一步的議價或下單動作，換言之，詢問是進入交易的第一步。

　　收到客戶的詢問後，須針對客戶所詢問的內容迅速回覆，以免客戶失去和你合作的耐心。回覆客戶時要詳盡，回覆的重點在於使對方對你的公司有信心並誘發其購買意願。一般而言，撰寫回覆詢問信函時，第一段要提及這封信主要是回覆哪些詢問；第二段則是切入主題回覆；第三段則是做禮貌性的結束或提醒對方等待進一步的回覆。

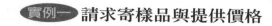
實例一 請求寄樣品與提供價格

Letterhead

Date: Mar. 10, 2013

ADDRESS

   Well received your full set catalogue. We are very interested in your products: A2332, A2433, A3323 , B5323 and C5633.

   I need further information on the **above-mentioned items**:

1. Please send two samples per each item.
2. Please **quote** us your best offer for the **above inquiry** based on **FOB Taiwan** as well as **CIF LA**.

   Please let me know **in advance** the amount if payment is required for the samples. **If you have any questions, please do not hesitate to let me know**.

Sincerely yours,

Tracy Chang

Tracy Chang

❖ 譯文

〜〜〜〜〜〜〜〜〜〜〜〜〜〜〜〜〜〜〜〜〜〜〜〜〜〜〜〜〜〜〜〜〜〜

　　已收到貴公司所提供的完整目錄。我們對以下的品號非常感興趣：A2332, A2433, A3323 , B5323 以及 C5633。

　　我需要上述品項的詳細資料：

1. 每項產品請各提供兩個樣品。

2. 請提供以上產品報價，報價基礎為 FOB 臺灣和 CIF 洛杉磯條件。

　　若樣品要收費，請事先告知。若有任何問題，請不要猶豫與我聯絡。

〜〜〜〜〜〜〜〜〜〜〜〜〜〜〜〜〜〜〜〜〜〜〜〜〜〜〜〜〜〜〜〜〜〜

相關詞彙說明

well received　已收到（是一種常見的表達法，可用 we
　received 來替代）

above-mentioned　以上所提；以上所述

item　品項（可用 product 取代）

quote　報價（可用 make an offer/offer your price 替代）

above inquiry　以上的詢問

base on　依據；以……為基礎

FOB Taiwan　臺灣港口船邊交貨條件（FOB = Free on Board）

CIF L.A.　洛杉磯成本運費保險含在內條件（CIF = Cost,
　Insurance, Freight）

in advance　事先；預先（可用 ahead/beforehand 取代）

If you have any questions, please do not hesitate to let me know.
若有任何疑問，請不要猶豫與我連絡。也可用以下的倒
裝句來表達 "Should you have any questions, please feel
free to let me know."

 詢問有關產品的詳細資料

Letterhead

Date: Mar. 13, 2013

Amerson Hardware Co., Ltd.
243 5th Ave. Manhattan
New York, NY 02234
USA

Att: Ms. Diane Chang,

Dear Diane,

Well received the samples from you. **After inspecting** the samples,
we have **following questions for you**:

1. Do you **go through electroplating process** at the **final procedures**?

2. Has the product gone through the following testing: **impact test, flattening test, persistent hydraulic test** and **dimensional test**?

3. If yes, please send us the **test report**.

4. If no, please explain the reason.

A prompt reply would greatly **facilitate** a quick **decision making**. **Thanks for your assistance**.

Sincerely yours,

Teresa Chen

Teresa Chen/Marketing Director

❖ 譯文

　　已收到貴公司所提供的樣品。在檢驗完後，我們有以下的問題要請教您：

1. 貴公司在最終製程時，有經過電鍍的程序嗎？

2. 產品是否有做以下的測試：撞擊測試、壓扁測試、持續水壓測試、尺寸測試？

3. 如果有經過以上測試的話，請提供各式報告。

4. 如果沒有的話，請說明理由。

　　請儘速回覆，以利敝公司做相關決定。謝謝您的協助。

相關詞彙說明

after inspecting　做過檢驗後（也可用 after inspection 取代）

following　以下的

questions for you　要向您請教的問題

go through　經過；歷經

electroplate　電鍍

process　過程；程序；步驟

final procedures　最終製程

impact test　撞擊測試

flattening test　壓扁測試

persistent　持續的；不間斷的

hydraulic test　水壓測試

dimensional test　尺寸測試（也可用 dimensional measurement 來取代）

test report　測試報告

facilitate　使容易；促進；幫助（也可用 help 取代）

decision making　做決定（可用 choice making 或 selection 替代）

Thanks for your assistance.　謝謝您的幫忙（也可用 Thanks for all your help. 替代）

實例三 詢問交易條件

Letterhead

Date: Mar. 14, 2013

Mr. Victor Chang/Marketing Mgr.
Ederson Electronics Inc..
243 8-Ks, Namdaemum-ro
Jung-Gu, Seuol
Korea

Dear Mr. Chang,

Thanks for the catalogue and 20 pcs of sample received last Tuesday. We are **fully impressed by** your catalogue and samples.

At this state, we are in the process of making final decision on our **seasonal purchase** for the **coming** summer. We need you to **provide us with** the following information:

1. Please send us your **price list** and **indicate** your **price terms**. Our **preferred price** terms is **C&F Le Havre.**
2. Please indicate your **delivery terms** and **lead time for** production.
3. Please advise your **payment terms**. Will **irrevocable** L/C at 30 days sight be acceptable?
4. Please indicate your packaging details **per item**.

5. Do you have any **discount scheme**?

6. What's your shipping terms?

Your **early response** in this regard would be very much appreciated.

Sincerely yours,

Elizabeth Hsie

Elizabeth Hsie

Purchasing Mgr.

❖ **譯文**

　　謝謝您寄來的目錄和 20 個樣品。我們在上週二就已收到這些東西。我們對您所寄來的目錄和樣品印象很深刻。

　　目前，我們正為即將來臨的夏天做季節性採購的最後決定。我們需要貴公司提供以下的訊息：

1. 請貴公司提供價格表。價格表上需註明價格條件。我們所希望的價格條件為 C&F Le Havre。

2. 請明示貴公司的交貨條件和生產所需期間。

3. 請告知貴公司的付款條件。不可撤銷 L/C 見票後 30 天付款的條件是否被接受？

4. 請說明貴公司每項產品的包裝細節。

5. 請問貴公司有沒有折扣方案？

6. 貴公司的運送條件為何？

　　如果貴公司能對以上幾點盡早回覆，我們將感激不盡。

相關詞彙說明

fully impressed by　對……留下非常深刻的印象

at this state　在這個狀況下（可用 under this circumstance 來取代）

seasonal purchase　季節性的採購

coming　即將到來的（可用 forthcoming/oncoming 來替代）

provide with + 名詞　提供……

price list　價格表

indicate　指明；指出；指示

price terms　價格條件

preferred price terms　所希望的價格條件（可用 favorable/favorite price terms 來取代）

C&F + 港口名稱　抵……港口成本運費含在內價格條件

delivery terms　交貨條件

lead time for　……的所需時間

payment terms　付款條件

irrevocable　不可取消的

per item　依照每一項產品（per 是每一或依照的意思）

discount scheme　折扣方案

early response　及早回覆（也可用 prompt/immediate /quick/ urgent reply）

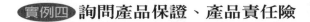
實例四 詢問產品保證、產品責任險

Letterhead

Date: Mar. 15, 2013

Mr. Raymond Ju/General Mgr.

Elite Hi-Tec. Inc.

243 West 17th Street,

Chicago, Il, 60600

USA

Dear Raymond,

Thanks for all the assistance that you have provided with **so far**. We **are about to place a bulk order with you**. However, we need to have your **final words** on the following **concerns**:

1. Do you have any **warranty** for your products? For how long does this warranty **last**?
2. Do you have any **Product Liability Insurance**? **If so**, Could you please **include** us **in** your **vendor list**?

It would be good if we could get your views on these questions

---

promptly.

Sincerely yours,

Maria Chen

Maria Chen

Purchasing Mgr.

❖ 譯文

〰〰〰〰〰〰〰〰〰〰〰〰〰〰〰〰〰〰〰〰〰〰〰〰〰〰〰〰〰〰

　　謝謝您到目前為止所提供的各種協助。我們即將要下一筆大單給貴公司。但是我們希望您對我們下列所關心的事提出決定的承諾：

　　1. 貴公司的產品有提供任何的擔保嗎？貴公司提供的保證期有多久？

　　2. 貴公司有提供產品責任保險嗎？如果有，貴公司可否將敝公司納入此保險的供應商清單中？

　　如果你可以盡早讓我們得知貴公司對以上問題的看法，那是最好也不過的事。

〰〰〰〰〰〰〰〰〰〰〰〰〰〰〰〰〰〰〰〰〰〰〰〰〰〰〰〰〰〰

相關詞彙說明

so far　到目前為止（可用 up until now/till now 來替代）

be about to + 加原型動詞　即將；正要

place an order with　下訂單給

a bulk order　大筆訂單（可用 considerable/substantial/large order 來取代）

final words　具決定性的承諾

concern　關心；掛念的事

warranty　保證；保證書

last　持續；維持

product liability insurance　產品責任險

if so　果真如此；若是如此（相反詞為 if not）

include in　將……包含在……內

vendor list　供應商清單（產品責任險中，一定使用到的專有名詞）

it would be good if...　如果……更好

實例五 詢問市場資訊

Letterhead

Date: Mar. 15, 2013

Advanced Pipes & Fittings Inc.
1100 Carlton Boulevard
Topeka, Kansas, 33443
U.S.A.

Att: Mr. Waller V. Walker/General Mgr.

Dear Waller,

Sub: Marketing Information

It is our great pleasure to learn that you express your interest in being our Sales Agent in Australia.

Before coming to any **final** reply, I would like to know your views on the following questions:

1. Are you a **retailer,** a **wholesaler**, a selling **agent** or a **Rept.**?
2. As we have **wide product lines** which **fit for** different **application**, please indicate your **major** activities are in **waste water treatment**, **chemical process**, chemical industry, plumbing, swimming pools, **water recreation**, **aquarium industry** or **aquatic farming**?
3. What is your market share in **local market**?
4. What are the **competitive products** you are encountering with?
5. Who are your major competitors? Are they **locally or globally based**?

We will be in touch with you right after receiving your **thought** to the above questions.

Sincerely yours,

Gertrude Lu
Gertrude Lu/G.M.

❖ 譯文

〰〰〰〰〰〰〰〰〰〰〰〰〰〰〰〰〰〰〰〰〰〰

主旨：行銷資訊

很榮幸得知貴公司有興趣成為敝公司在澳洲地區的銷售代理商。

在做決定性的回覆前，我們想知道您對以下問題的看法：

1. 貴公司的營業性質是零售商？批發商？業務代理商？或是業務代表？

2. 由於敝公司的產品線很廣，且其應用範圍也不同，請說明貴公司的主要業務是適用於廢水處理？化學製程？化工業？民生用水管？游泳池工業？水上休閒娛樂業？水族館業？或是水產養殖業？

3. 貴公司在當地市場的佔有率是多少？

4. 貴公司所面臨的競爭性產品有哪些？

5. 貴公司主要的競爭者是誰？這些競爭者是以當地或是以國際為根據地？

在得知貴公司對以上問題的看法後，我們將與您聯絡。

〰〰〰〰〰〰〰〰〰〰〰〰〰〰〰〰〰〰〰〰〰〰

相關詞彙說明

final　原指最後的、最終的，這裡引申為具決定性的。

retailer　零售商；零售店

wholesaler　批發商

agent　代理商

Rept.　代表（是 representative 的縮寫）

wide product line　廣大的產品線

fit for　適合……

application　應用領域

waste water treatment　廢水處理

chemical process　化學製程

water recreation　水上休閒娛樂

aquarium industry　水族業

aquatic farming　水產養殖業

local market　當地市場

competitive products　競爭性產品

encountering with　面臨（可用 face 來替代）

locally or globally based　以當地或以國際為根據地

thought　原指思想，這裡引申為觀點和看法之意。

 「請求寄樣品與提供價格」的回覆

Letterhead

Date: Mar. 12, 2013

ADDRESS

Dear Tracy ,

    Thanks for your letter dd. Mar. 10 **with respect to** your **request of** sample mailing and best offer for the following items: A2332, A2433, A3323 , B5323 and C5633.

    Two samples per each items were sent today **via** UPS. The **tracking no.** is WT100 223 233. The total amount for the request samples including the freight cost is US$150. Please **issue** a **cashier's check** per the attached **invoice** and mail it to us. **As for** the **pricing** for the **requested** items, please **have a look at** the attached **quotation sheet**.

    I hope you will be satisfied with our samples and quotation. Please confirm receipt of all the samples and quotation **by return**. Thanks!

Sincerely yours,

Tracy Chang

Tracy Chang

❖ **譯文**

〰〰〰〰〰〰〰〰〰〰〰〰〰〰〰〰〰〰〰〰〰〰〰〰〰〰〰〰〰

　　謝謝貴公司三月十日來函要求有關寄送 A2332, A2433, A3323, B5323 以及 C5633 等產品的樣品，和我方最優惠價格。

　　我們已於今天將每款各兩個樣品以 UPS 寄出。提單的跟 號碼為 WT100 223 233。這些樣品包含運費的費用總計為 US$150。請依據所附上的發票開一張銀行本票寄送給敝公司。至於貴公司對所要求的品項報價部分，請參考附件的報價表。

　　希望貴公司能對我們的樣品和報價感到滿意！在收到樣品和報價後，請回覆並確認。謝謝！

〰〰〰〰〰〰〰〰〰〰〰〰〰〰〰〰〰〰〰〰〰〰〰〰〰〰〰〰〰

相關詞彙說明

with respect to　與……有關（也可用 with reference to/with regard to/with concerns to/regarding/concerning 等用語來取代）

request of　……的要求（也可用 requirement of）

via　透過；藉由（也可用 by）

tracking no.　跟催號碼；追蹤號碼

issue　開立

cashier's check　銀行開具的本票，與 personal check（個人開據的本票）不同，因為 cashier's check 一定保證兌現；而 personal check 則不一定保證兌現；另外還有 certified check（保兌支票）。

invoice　發票

as for　至於（也可用 as to/as regards）

pricing　訂價；價格

requested　所要求的（也可用 required）

have a look at　看；參照

quotation sheet　報價單

by return　回覆；回報

 實例七　「詢問有關產品的詳細資料」的回覆

Letterhead

Date: Mar. 14, 2013

EtexGlyn Inc.

630 Cutter Mills Street,

Great Neck, NY 11021

USA

Att: Ms. Amanda Cheng/General Manager,

Dear Amanda,

Thank you very much for your letter of March 13 in which you

---

confirmed receipt of the samples. You also indicated some questions in this letter. The following are the replies to your concerns:

1. We **do** go through an electroplating process at the final procedures.
2. After finishing the production, all products must go through the following tests: impact test, flattening test, **bursting test**, persistent hydraulic test, dimensional measurement such as **ID, OD, wall thickness** and **roundness**.
3. We will send you the test report separately **by post**. However, not all the test reports are **available** for **confidential purposes**.

I hope that the above replies can **answer all your concerns**. If you need further information, **you are always the most welcome to contact us.**

Sincerely yours,

Diane Chang

Diane Chang/General Mgr.

❖ 譯文

非常謝謝您 3 月 13 日的來信，您在信中確認已收到了樣品，並指出了一些問題。以下是我們對您提出的疑慮所做出的回覆：

1. 我們在最終製程中確實有電鍍的程序。

2. 在產品製作完成後，所有的產品都會做以下的測試：撞擊測試、壓扁測試、爆破測試、持續水壓測試、尺寸測試諸如內徑、外徑、壁厚和真圓度的量測等。

3. 我們會另外以郵寄方式將測試報告提供給貴公司。但是，基於機密考量，並非所有的報告都可提供給貴公司。

希望以上已完全回覆了您的疑慮。如果您需要更多的訊息，歡迎您與我們聯絡。

相關詞彙說明

do　確實、的確之意，用來加強語氣（也可用 indeed）

bursting test　爆破測試

ID.　內徑（是 inside diameter 的縮寫）

OD.　外徑（是 outside diameter 縮寫）

wall thickness　壁厚

roundness　真圓度

by post　以郵寄的方式

available　可取得；可取用

confidential purposes　機密原因

answer all your concerns　回覆您所有的疑慮

You are always the most welcome to contact us.　隨時歡迎您
　　與我們聯絡。（可用 please feel free to contact us/please
　　do not hesitate to let us know 等用語來取代。）

 實例八　「詢問交易條件」的回覆

Letterhead

Date: Mar. 16, 2013

Ms. Elizabeth Hsie/ Purchasing Mgr.

Korea Telecommunication Inc.

888 7-Au, Namdaemum-ro

Jung-Gu, Pu-Sang

Korea

Dear Ms. Hsie,

Allow me to begin by thanking you for confirming receipt of the catalogue and samples.

The following are the responses to your inquiries :

1. The requested price list is as attached. We have quotation based on

two different terms: FOB Taiwan (our standard quotation) and C&F Le Havre (your preferred one).

2. Our standard delivery terms is 30 days. The lead time for production takes 20 days. This means that we need 10 days for shipping arrangement.

3. We accept payment terms of T/T in advance, **T/T against document**, and Irrevocable L/C at sight. As this is the trial order for we both, sight L/C at 30 days **is** not **considered**.

4. We do have discount scheme. We offer 1~3% **annual rebate** based on your annual **purchasing volume**. Please see the attached detailed sheet.

5. We can ship **either by ocean or by air freight. Unless** the customer has his own shipping forwarder, we **normally** use our **appointed forwarder**.

I hope I have replied to all your questions. If not, feel free to let me know.

Sincerely yours,

Victor Chang

Victor Chang

General Manager

❖ 譯文

　　首先謝謝您確認已收到目錄和樣品。以下是針對您所提問題的相關回覆：

1. 您所要求的報價已附上。我們提供了兩種報價：其一為敝公司的標準價FOB Taiwan；其二為您所想要的C&F Le Havre。

2. 我們的標準交貨期為30天，生產期要20天。這表示我們需要10天的時間做出貨安排。

3. 我們接受的付款條件為出貨前電匯、憑單據電匯、以及即期信用狀。因為此次交易對雙方而言是嘗試性的訂單，信用狀見票30天後付款的條件，不在考量的範圍內。

4. 我們確實有折扣貼現方案。我們依貴公司年度的採購金額提供1~3%不等的折扣貼現。請看附件細節說明。

5. 我們以海運或空運寄送貨品。除非客戶有自己指定的貨運承攬人，否則我們會使用敝公司指定的貨運承攬人。

　　希望我已針對您所有的疑問做出回覆。如果有回覆不周之處，請讓我知道。

相關詞彙說明

T/T against document　憑單據電匯付款

be considered　做為考量（名詞可用 be under consideration 來表達）

annual rebate　年度折扣貼現

purchasing volume　採購量（purchase 可當名詞和動詞。這裡的 purchasing 是動名詞當形容詞來用；而 volume 是指量而言，可指數量或金額量）

either ... or...　不是……就是……

by ocean　以海運的方式（可用 by sea 來取代）

by air freight　以空運的方式（可用 by air/by air shipment 來替代）

unless　除非（也可用 except/only if 來替代）

normally　通常；按慣例（也可用 usually/often 來取代）

appointed forwarder　指定的貨運承攬業者（可用 assigned forwarder 來取代）

Letterhead

Date: Mar. 18, 2013

Ms. Maria Chen/Purchasing Mgr.

Advanced Hi-Tec. Inc.

10F, Nou Plaze, Jalan P Rame

Kuala Lumpur

Malaysia

Dear Maria,

Your urgent request with respect to a bulk order has been received. The following are my final words to your concerns:

1. We do have product warranty which lasts for 2 years from **shipping invoice date**.

2. We do have product liability insurance as well. However, we do not include all our customers in the vendor list. Only those customers whose major activities are in chemical process, **public works**, waste water treatment, chemical industry are included in the vendor list. **In addition to** the above conditions, customers whose annual

---

purchasing volume **exceed** US$300,000 are also included in this **insurance policy**.

After **checking** your **records**, we feel sorry to find out that you do not meet the requirements mentioned as above. **It will not be possible to accommodate your request for the vendor list.**

Your understanding on this matter is highly appreciated. Looking forward to hearing from you soon about the bulk order.

Sincerely yours,

Raymond Ju

Raymond Ju

General Manager

❖ **譯文**

　　已收到貴公司有關緊急大單的要求。以下是我對您的疑慮所做出的最後承諾：

　　1. 敝公司確實有自出貨發票日算起，為期兩年的保證期限。

　　2. 我們確實也有產品責任險。但是我們並未將所有的客戶納入保險清冊中。只有那些主要商業領域是化學製程、公共工程、廢水處理、化工業者才會納入這份保險中。除了以上的條件外，年採購額達美金三十萬的客戶也會列入我們的保險單中。

在查過貴公司的記錄後，我們很遺憾的發現貴公司並未符合以上所提的要求。我們很難如您所願，將貴公司納入產品責任險的客戶清單中。

貴公司對這件事的體諒我們將感激不盡。期待早日收到貴公司的大量訂單。

〜〜〜〜〜〜〜〜〜〜〜〜〜〜〜〜〜〜〜〜〜〜〜〜〜〜

相關詞彙說明

shipping invoice date　出貨發票日

public works　公共工程；公共建設工程

in addition to　除……以外（也可用 apart from 來取代）

exceed　超過（可用 be more than 來替代）

insurance policy　保險方案

check record　核對資料（也可用 review records 來替代）

It will not be possible to accommodate your request for the
vendor list.　我們很難如您所願將貴公司納入在保險客戶
清冊中（這是一種委婉的拒絕方式，比直接告訴客戶 We
can not accept .... 或 We are not able .... 來得客氣委婉）

Your understanding on this matter is highly appreciated.　貴公
司對這件事的體諒我們將感激不盡（也可用 "Your
understanding on this matter is highly regarded." 來替
代）。

詢價、報價、議價

Pricing Inquiry, Quotation,
Counter Offer

　　在客戶對產品、交易條件和其他相關的問題有一定的了解後，就相當接近下單的階段。可是在下訂單前，有一個過程是非常重要且不可或缺的，那就是詢價、報價和議價。

　　詢價 (Pricing Inquiry/Quotation Inquiry)，顧名思義就是指買方針對有興趣的產品對賣方所做的價格詢問，又稱 Buyer's Offer 或 Buying Offer。報價 (Quotation/Offer) 是指賣方針對買方的詢價所報出的價格，所以又稱 Seller's Offer 或 Selling Offer。議價 (Counter Offer) 是買賣雙方針對價格討價還價的過程。以買方的立場來看，當然價格是越低越好；而以賣方的立場來看，則是要顧及成本和利潤，所以議價是必然的情況。這時不論是買方或賣方，就必須要有足夠讓人信服的說法讓對方接受各方認為合理的價格區間。

 **詢價**

Letterhead

Date: Mar. 15, 2013

Ms. Pamela Hsu

FIC Manufacturer Inc.

566 Vineyard Ave.

Singapore 1124

Dear Pamela,

Sub: Quotation for Inquiry for Dual Line Extruder, Your Model #KK555

A customer of ours is interested in your Model#KK555, dual line extruder for **tubes** for gas application. Please send us your **best offer** based on the following information:

Qty: 2 sets

Price terms: CIF Kaohsiung and FOB Singapore

The customer will **place a trial order with** you if your **pricing** is **competitive**. Please also indicate the delivery time in your quota-

tion sheet.

Looking forward to your kind reply.

Sincerely yours,

Monica Jen

Monica Jen

❖ 譯文

主旨：雙生線押出機，貴公司型號 KK555 之報價

　　敝公司的一位客戶對貴公司的KK555型號，用以生產瓦斯管專用的雙生產線機器非常感興趣。請依據以下的資訊，提供貴公司最優惠的報價：

　　數量：2

　　價格條件：CIF 高雄和 FOB 新加坡

　　如果貴公司的價格具競爭性，這位客戶將下試驗性的訂單給貴公司。請在貴公司的報價單中說明交貨期間。

　　期待收到貴公司善意的回覆。

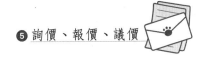

相關詞彙說明

dual line　雙條生產線（dual 可用 double 來取代；line 原指線，這裡引申為生產線。

extruder　押出機；擠出機

tubes　管

best offer　最好的報價（可用 best price/the most competitive price 來替代）

place an order with　下訂單給（可用 give an order/grant an order）

a trial order　試驗性的訂單

pricing　定價（price 當名詞時，指價格而言；price 當動詞時，是定價的意思。這裡將 pricing 動名詞當作名詞使用）

competitive　具競爭 的（若用來形容價格具競爭性時，可用 reasonable/good/the best 來形容）

quotation sheet　報價單（可單用 quotation/price 來取代）

實例二 量產詢價

Letterhead

Date: Mar. 16, 2013

Ms. Branda Hsu

Branda Piping System Co.

223 Queen Street

S-207 46 Stockholm

Sweden

Dear Branda,

Sub: Soliciting bulk inquiries

As the economic in Asia is **booming** after the war, there is a lot of **construction** in the major cities in Asia. Our company recently have **won** several **bids for fire sprinkler systems** in office building in Beijing, Tokyo, Shanghai, Seoul and Taipei.

We have bulk inquiry for the following items:

1/2" ~4" ASTM Sch. 80 pipes 450000 M

1/2" ~4" ASTM Sch. 80 fittings 25000 per each fitting

Primer and **cement** 60000 cans (2 pint s/can)

Delivery terms: May 1ˢᵗ ~ July 31, 2013

Price terms: CIF Taiwan

Payment terms: T/T 10 days after from invoicing date

If your prices are reasonable, we will place the orders **without any delay**.

Sincerely yours,

Stephanie Boo

Stephanie Boo

General Manager

❖ 譯文

主旨：量產詢價

　　隨著亞洲經濟戰後日益復甦，亞洲的主要大城市出現許多建設工程。敝公司最近在北京、東京、上海、漢城和臺北等城市，贏得了許多辦公大樓消防灑水系統的標案。

　　我們欲對以下產品做量產詢價：

1/2" ~4" ASTM Sch. 80 管 450000 公尺

1/2" ~4" ASTM Sch. 80 管配件 25000 每一種配件

清潔劑與膠水 60000 罐（2 品特 / 罐）

交期：2013 年 5/1～7/31 期間

價格條件：成本、運費、含在內臺灣港口價

付款條件：發票日後十天電匯付款

如果貴公司價格合理，我們會馬上下單。

〰〰〰〰〰〰〰〰〰〰〰〰〰〰〰〰〰〰〰

相關詞彙說明

soliciting　原型為 solicit，是請求的意思

booming　復甦；景氣好（可用 is getting better/improving 來取代）

construction　建設工程

win bids for　贏得……標案

fire sprinkler systems　消防灑水系統

primer　清潔劑

cement　膠水

without any delay　隨即（可用 immediately/promptly/quickly 等取代）

實例三 報價(一)

Letterhead

Date: Mar. 18, 2013

Ms. Monica Jen
PP Tube Manufacturing Corp.
No. 999, Ming Tsu 1st Rd., San Ming Dist.
Kaohsiung, Taiwan, R. O. C.

Dear Monica,

Sub: Quotation for Dual Line Extruder, our Model #KK555

Thanks for your inquiry on our Model # KK555. We are pleased to provide you with our best offer as following:

Model #KK555: Extruder for PP Tubes
1 dual line extruder, **Conical screws with molybdenum-plated, 4 barrel zones**, 2 **moulds heads**, dual line **cooling systems**, 1 **downstream equipment**, 2 **cutters**, 2 **haulers**, one **auto-printing machine.**
Price: US$ 5,550,000/set FOB Singapore
Price: US$556,500/ set CIF Kaohsiung

Delivery terms: one to be shipped 2 months after the confirmation on P/I; the other to be shipped 3 months after the confirmation on P/I. Payment terms: Irrevocable L/C at sight.

As we have received considerable orders since this season. We would like to **urge** you to confirm this order promptly. **That way**, we can arrange the production to meet **the said** delivery terms.

Let me know if you have further questions.

Sincerely yours,

Pamela Hsu

Pamela Hsu

❖ 譯文

主旨：敝公司型號 KK555 雙線押出機的報價

謝謝貴公司對敝公司產品型號 KK555 的詢價。我們很高興提供如下的優惠價格給貴公司：

產品型號 KK555: PP 管押出機：

一部雙線生產線押出機、鍍鉬圓錐形螺桿、四個套筒區、兩副模頭、雙線冷卻系統、一個下游設備、兩個切割機、兩個牽引機、一個自動噴印機。

價格：US\$ 5,550,000 ／每部 FOB 新加坡

價格：US\$556,500 ／每部 CIF 高雄

交期：確認銷貨確認書後的兩個月出貨一部；另一部則是在
確認銷貨確認書後的三個月出貨。

付款條件：不可撤銷即期信用狀。

因為我們從這季開始已收到了大量的訂單，所以想敦促貴公司盡快確認此訂單。這樣我們就可趕快安排生產，以符合上述的交期。

若有任何疑問，請讓我知道。

相關詞彙說明

conical screw　圓錐形螺桿

with molybdenum-plated　鍍鉬的

barrel zone　套筒區

mould head　模頭

cooling system　冷卻系統

downstream equipment　下游設備

cutter　切割機

hauler　牽引機

auto-printing machine　自動噴印機

urge　促；敦促

that way　那樣的話（可用 by doing that 來取代）

the said　上述的（可用 The above mentioned/the above 來取代）

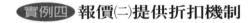 報價㈡提供折扣機制

Letterhead

Date: Mar. 18, 2013

Ms. Stephanie Boo

SB Engineering Consultancy Inc.

No. 666, Cheng Kuo North Rd.

Taipei, Taiwan, R.O.C.

Dear Ms. Boo,

Sub: Rebate scheme

Well received your large inquiries on pipes, fittings, primer and cement. I think you **must have** received the pricing by e-mail from my secretary yesterday. Now, I would like to **revert to you on** the discount scheme.

As your inquiry is in deed great quantities, you deserve the following rebate:

Volume/Order	rebate
US$100,000	1%
US$300,000	2%
US$450,000	3%
US$600,000 and **above**	4%

You can **ask for refund** or for **credit note** for the rebate that you gain per order. This is very attractive rebate scheme. We hope that this can help you to make quick decision. Once the order is confirmed, we need you to give us a **pre-advice two weeks ahead** your placing an order for we need to prepare the materials for this big order.

Looking forward to a **mutual profitable** business relationship with you.

Sincerely yours,

Branda Hsu

Branda Hsu

General Manager

❖ 譯文

主旨：折扣機制

　　已收到貴公司在管材、管配件、清潔劑和膠水上的量產詢價。我想貴公司已收到我祕書的報價電子郵件，現在我想針對折扣計劃做相關答覆：

　　因為貴公司的詢價量確實很大，所以值得獲得以下的折扣貼現：

採購金額／訂單	折扣貼現
US$100,000	1%
US$300,000	2%
US$450,000	3%
US$600,000 (含)以上	4%

　　貴公司在每張訂單獲得的回饋金，可以要求退款或開折讓單，這是很具吸引力的折扣貼現方案。我們希望這樣的方案有助於貴公司快速做出決定。一旦訂單確認後，我們希望您在下單的兩週前預先通知我們，因為我們需要準備這張訂單的原料。

　　期待與貴公司建立一個雙方互利的商業關係。

相關詞彙說明

must have + 過去分詞　必定……（對過去的事實做一假設
　　或猜測）

revert to 人 on...　針對……回覆某人

above　高於；以上（可用 more than）

ask for　要求；請求

refund　退款

credit note　折讓單

pre-advice　事前通知

two weeks ahead　兩週前（可用 two weeks before/two weeks
　　in advance）

mutual profitable　雙方互利（可用 reciprocally profitable 替代）

實例五 議價㈠因報價高於市價、競爭者而要求降價

Letterhead

Date: Mar. 22, 2013

Yogurt Maker Inc.
1100 Carlton Boulevard
Topeka, Kansas, 33443
U.S.A.

Att: Ms. Leslie Wu./Marketing Director

Dear Leslie,

Sub: More Competitive Prices

Well received your quotation on **yogurt** in strawberry, blueberry, vanilla, pineapple and tangerine **flavors** two weeks ago.

Our marketing team has **compared** your pricing **with** market pricing and other competitors' offer. We feel sorry to find out that your pricing is 3% higher than the market price, and 2% higher than the competitors' as well. We have **test** your product **in the market** and the **consumers** are very satisfied with the **taste**. The only thing they are not satisfied is your pricing.

We see a **potential market** for your products, especially, we have strong **distributing net** in all regions. Our **target price** is to **lower down** 6% than the previous quotation. If you consider the future **big demand**, you will agree that 6% discount is **worthwhile**.

Your good intention is highly expected.

Sincerely yours,

Pauline Chen
Pauline Chen

❖ 譯文

主旨：更具競爭性的價格

　　已在兩週前收到貴公司草莓、藍莓、香草、鳳梨、橘子口味優格的報價。

　　敝公司的行銷團隊將貴公司的報價跟市價與其他競爭廠商的報價相較之後，很遺憾得知貴公司的報價比市價高出 3%，較競爭者高出 2%。我們已將貴公司的產品在市場上做測試，且消費者相當滿意貴公司的口味。他們唯一不滿意的是貴公司的定價。

　　我們看到貴公司的產品頗具市場潛力，尤其是我們在各區域有很強的銷售網路。我們的目標價是將貴公司先前的報價往下調整 6%。如果貴公司考量到未來的廣大需求，貴公司會同意調降 6%。

　　期待貴公司的善意回覆。

相關詞彙說明

yogurt　優格

flavor　口味

compare with　與……相比較

test in the market　在市場上做測試

consumer　消費者

taste　口味（可用 flavor 替代）

potential market　市場潛力

distributing net　銷售網路

target price　目標價（可用 ideal price/favorable price 替代）

lower down　往下調降

big dimand　廣大需求

worthwhile　值得的（可用 worthy 替代）

Your good intention is highly expected.　我們期待貴公司的善
　　意回覆。（這是非常制式的客氣用語，用來表達期待有
　　善意的回覆）

議價㈡因市場價格滑落而要求降價

Letterhead

Date: Mar. 22, 2013

Card Printing Co., Ltd.

3 Cleveland Street,

London SW#B 5BD

ENGLAND, U.K.

Att: Ms. Winifred Yang/Marketing Director

Dear Ms. Yang

Sub: Marking down the pricing

We have received your price list and have **looked over** it. We feel sorry that your price is more expensive than the **inside price** for this **article** and it's **above our limit**.

There was **a fall in the paper pulp price** last month, and this is **known to everyone**. Maybe you have not yet **reflected** this **drops** when you **set the price**. If so, we would like to ask you to **re-quote**. There is a tendency for **a slumping in the foreseeable future**. We hope that you can consider this in your future price. If your offer meets our ideal price, we will place the order with you next month.

Your good intention is highly expected.

Sincerely yours,

Fenny Wu
Fenny Wu

❖ 譯文

〰〰〰〰〰〰〰〰〰〰〰〰〰〰〰〰〰〰〰〰〰〰〰

主旨：降價

　　我們已收到貴公司的報價，且很仔細地看過這份報價。我們很遺憾發現貴公司的價格較國內市場的價格來得高，超出了我們的界限。

　　如眾所知，上個月紙漿有價格下跌的現象，或許貴公司在定價時尚未反映這部分，若是如此，我們想請求貴公司重新報價。在可預見的未來，價格將會有暴跌的趨勢，我們希望貴公司未來定價時能將此點列入考量。如果貴公司的報價符合我們的理想價，我們將於下個月下訂單。

　　期待收到貴公司的善意回覆

〰〰〰〰〰〰〰〰〰〰〰〰〰〰〰〰〰〰〰〰〰〰〰

相關詞彙說明

mark down　降價（也可用 cut down/reduce/knock down/
　　decrease/diminish/level down/lower down/drop 取代）

look over　仔細檢查

inside price　國內價

article　產品；品項（可用 product/item 替代）

above our limit　超出我們的界限（可用 further to our limit；
　　far from our expectation 替代）

a fall in price　價格上的滑落

paper pulp 紙漿

known to everyone 眾所皆知

reflect 反映

drop 降落（可用 fall 取代）

set a price 定價（可用 make/fix/publish/list a price 取代）

re-quote 重新報價（可用 re-offer 替代）

a slumping 暴跌（可用 collapse 替代）

in the foreseeable future 在可預見的未來

實例七 議價㈢同意降價

Letterhead

Date: Mar. 21, 2013

Dolls Manufacturer Inc.

3 Cleveland Street,

London SW#B 5BD

ENGLAND, U.K.

Att: Mrs. Roselyn Lin/Sales Manager

Dear Mrs. Lin,

Sub: Breaking down the pricing

Well received your letter dd. March 10 with respect to your target price. Please be advised that we have **shaved our price as far as possible** and further reduction may **not be possible without sacrificing quality**.

However, **in view of** our **mutually-satisfactory** relationship **to date**, there is **nothing worth** going to **war about on** the 1% discount. Therefore, we have made up our mind to **concede** a further 1% discount. This price will **be effective retroactive to** February.

We hope that you could be satisfied with this decision. **We look forward to receiving your orders as per your promise**.

Sincerely yours,

Yvonne Cheung

Yvonne Cheung

❖ 譯文

主旨：降價

　　已收到貴公司三月十日與貴公司目標價有關的來函。我們想告知貴公司我們已盡可能的削減敝公司的價格，若再進一步的調降將無法避免的犧牲品質。

　　然而，有鑑於雙方迄今的圓滿關係，實在不值得為 1 ％的降價爭論不休，所以我們已決定進一步調降 1 ％，新價的有效性將追溯自二月起。

　　我們希望貴公司能對此決定感到滿意。我們期待收到如貴公司所承諾的訂單。

相關詞彙說明

break down the price　降價

shave price as far as possible　盡可能的削減價格（可用 do our best to shave off the price 替代）

not be possible without　無法避免不……（without 後面的動詞只能用動名詞的形式，不可加原型 詞）

sacrifice quality　犧牲品質

in view of　有鑑於（可用 in consideration of 替代）

mutually-satisfactory　雙方滿意的

to date　到今天為止

nothing worth + Ving　不值得……（worth 後面的動詞只能用動名詞的形式，不可加原型動詞）

war about on　在……上爭論（可用 argue about on/dispute on 取代）

concede　給予；容許

be effective　有效

retroactive to　追溯自

We look forward to receving your orders as per your promise.
我們期待收到如貴公司所承諾的訂單。（可用 "We hope to get the orders as per your promise." 替代）

實例八 議價㈣拒絕降價──按公司政策辦事，原報價仍較具競爭力

Letterhead

Date： Mar. 19, 2013

Int'l Hardware Co., Ltd.

666 Queen Elizabeth Street

Brecksville, IL 22343

USA

Att: Ms. Jacqueline Roselyn Lin/Marketing Mgr.

Dear Jacqueline,

Well received your letter dd. March 10 with respect to your request in breaking down the prices.

Please understand that we are always interested in **supporting your sales efforts in every way we can**. The prices we quoted **are firm** and **consistent with** our pricing for all markets. **For this very reason**, it will **not be possible to accommodate your request for** a special reduction.

You are strongly **encouraged** to reflect on the **outstanding quality** of these products and the **commanding position** they have in your highly competitive market. There is **very reason to believe** that it will **do at least as well** in your market.

We look forward to the opportunity of doing business with you.

Sincerely yours,

Melisa Wu
Melisa Wu

❖ **譯文**

〰〰〰〰〰〰〰〰〰〰〰〰〰〰〰〰〰〰〰〰〰〰〰

　　已收到貴公司三月十號有關降價的要求。

　　希望貴公司了解我們總是盡可能的在各方面支持貴公司對業績所做出的努力。我們所提供給貴公司的報價是固定的，且與其他市場的價格一致。有鑑於此，要敝公司接受貴公司的特別價格的要求是有困難的。

　　我們建議貴公司在市場上強調敝公司優異的產品品質，並強調敝公司產品在高度競爭的市場上所佔有的主導地位。我們有充分的理由相信這樣的價格在貴公司的市場中至少可做得不錯。

　　我們期待有機會與貴公司合作！

〰〰〰〰〰〰〰〰〰〰〰〰〰〰〰〰〰〰〰〰〰〰〰

相關詞彙說明

support your sales effort　支持貴公司的業績努力

in every way we can　盡我們所有的可能（可用 do our utmost
　　effort 替代）

be firm　固定的（可用 be fixed 替代）

be consistent with　與……一致（可用 in accordance with/in
　　conformity with/be identical with 替代）

for this very reason　正是這個理由（very 在這裡是指恰好／
　　正好是的意思）

not be possible to accommodate your request for　無法符合您
　　在……的請求

encourage　原意是鼓勵，這裡引申為建議（可用 recommand 替代）

outstanding quality　優異的品質

commanding position　主導的地位（也可用 leading position 替代）

very reason to believe　充分的理由相信（very 在這是指充分的／完全的意思）

do at leaset as well　至少可做得不錯

We look forward to the opportunity of doing business with you.　我們希望有與貴公司合作的機會。（是一種常用的結尾用語）

實例九 議價㈤拒絕降價——但提出替代品方案

Letterhead

Date: Mar. 18, 2013

Coffee Machine Industrial Cor.
666 Queen Elizabeth Street
Brecksville, IL 22343
USA

Att: Ms. Lorry Wu/General Mgr.

Dear Ms. Wu,

Thank you for your counteroffer of March. 15 in which you request a **workable price** in your market for our Coffee Machines.

Please understand that we **fully realize** the importance of providing with quality products **under conditions** that will permit you to win a **commanding share** in your market. This is **the spirit** in which our original offer was made.

We do appreciate your position and, therefore, would like to offer **an alternative**. You can order our model B series which we can **fix the prices at your limit**. This series should **be more than adequate for** your requirements. Enclosed is additional information on this series.

Please let us have your final thought on which model you **decide to go with**.

Sincerely yours,

Jessica Fang

Jessica Fang

❖ 譯文

〜〜〜〜〜〜〜〜〜〜〜〜〜〜〜〜〜〜〜〜〜〜〜〜〜〜〜〜〜

　　非常謝謝貴公司在三月十五日的議價信，貴公司在此信中要求敝公司提供貴公司可行的咖啡機價格。

　　我們完全了解在提供貴公司高品質產品的前提下，讓貴公司在市場上贏得領導性佔有率的重要性。這也是我們原始報價的主要用意。

　　我們明白貴公司的立場，所以想提供另一替代方案。貴公司可訂購敝公司的 B 系列產品，敝公司可將此系列產品價格定在貴公司可接受的範圍內，且這系列產品應該更符合貴公司的需求。附件是這系列產品的額外資料。

　　請告知貴公司最後決定要採用哪一種產品。

〜〜〜〜〜〜〜〜〜〜〜〜〜〜〜〜〜〜〜〜〜〜〜〜〜〜〜〜〜

相關詞彙說明

workable price　可行的價格

fully realize　十分瞭解（可用 completely realize 替代）

under conditions　在……的前提、條件下

commanding share　領導 的市場佔有率

the spirit　原指精神之意，這裡引申為用意。

an alternative　替代方案（可用 another workable way 替代）

fix the prices at your limit　將價格定在你可接受的範圍內

be more than adequate for　更合適；更恰當

decide to go with　決定採行

 議價㈥調高價格

Letterhead

Date: Mar. 19 2013

Sponge Industrial Co., Ltd.

666 Queen Elizabeth Street

Brecksville, Segovia

Spain

Att: Ms. Olive Yang/Marketing Mgr.

Dear Olive,

Allow me to begin by thanking for all you have done to **secure a foothold** for our products in your market in spite of the **negative factors affect on** your business. Your ability to **maintain good sales** in the past years is **highly regarded**.

We have also had our difficulties. The biggest of these has been the increasing of **material cost**. The price for the material has been **going upward** since two month ago, and it's continuously advancing. In addition, the **sharp depreciation** in the value of Dollar on which

all our prices are based is another factor. We have **done every effort** to absorb this drop in revenue. Unfortunately, the depreciation now is reaching 10% and is still moving upward. It has become impossible to **unilaterally absorb** the loss, especially we have been **running in red** since one month ago. **Consequently**, we are **reluctantly compelled** to raise our export prices 5% **across the board** effective April 1st. We acknowledge that this will not make your situation easier. However, **bear in mind** that we are only asking for 5%. We are not seeking to **shift** the **entire burden** to you.

Your understanding and cooperation in this regard will be highly appreciated.

Sincerely yours,

Sabrina Chen

Sabrina Chen

❖ 譯文

〰〰〰〰〰〰〰〰〰〰〰〰〰〰〰〰〰〰〰〰〰〰〰〰〰〰

　　首先謝謝貴公司所做的一切。儘管眾多不利因素影響貴公司的生意，但貴公司仍努力使敝公司產品在貴市場立於穩固之地。貴公司過去幾年也來一直維持良好的銷售成績，您的努力我們相當的重視。

　　我們也有我們的難處。其中尤以原料成本價的增加為最大的難題。原料成本價格在兩個月前就已向上攀升，而且還在持續攀升中。此外我們計價的基礎——美金劇烈貶值也是另一因素。我方已盡全力從盈收吸收損失。不幸的是，目前的貶值幅度已達10％，而且數字還在繼續往上攀爬。我們已很難單方面吸收這個損失，尤其我們在一個月前營運就開始出現赤字。因此，我們被迫自四月一日起，全面將價格提高５％。我們了解這樣做會讓貴公司的處境更艱辛，但是請記住我們只要求提高５％，而並未要求將所有的負擔轉嫁到貴公司身上。

　　貴公司在這點的體諒與合作，敝公司將感激不盡。

相關詞彙說明

secure a foothold　穩固據點；穩固立足點

negative factors　負面因素

affect on　對……的影響

maintain good sales　維持好的業績

highly regard　高度重視（相當於 highly value）

material cost　原料成本

going upward　向上攀升；持續攀爬

sharp depreciation　急劇貶值（可用 dramatical depreciation）

all our prices are based　我們所有價格的計價基礎

do every effort　盡了所有的努力

unilaterally absorb　單方面吸收

run in red　營業出現赤字

consequently　結果；因此（可用 as a result/therefore 取代）

reluctantly compelled　被迫

across the board　全面；一律

bear in mind　記住

shift　原是轉移的意思，這裡引申為轉嫁（可用 transfer/
　　　move to 取代）

entire burden　所有的負擔

下訂單、修改訂單

Placing/Revising the Orders

　　前一章所提的詢價、報價、議價，都是為了要進入這一章的主題——下訂單而必須做的準備。換言之，若買賣雙方對前兩章所提的各相關條件都滿意，則就可以進行買賣交易，而這正是整個貿易流程或開發信的核心重點——促使對方進行下單的動作。必須謹記的是，買賣交易可能會因雙方難以預料的情況而有所變，面對這樣的情況時，必須本著服務客戶的心耐心以對，並盡量符合客戶的要求，才能成為商場的常勝軍。

　　以下是下單或修改訂單常用到的英文用語：

(1)下訂單：place an order with; make an order; pass an order; send an order; file an order; give an order; grant an order; favor an order 不過後面的三種說法較不適合下單者使用，否則易給人驕傲的感覺。若是由賣方所使用，則給人客氣的感覺。

(2)訂貨單：order shee; indent

(3)初始訂單：initial order; first order

(4)試驗性訂單：trial order

(5)生產訂單：execute an order; carry out an order; fulfill an order; execution of an order; proceed an order

(6)促使下單：canvass for an order; induce for an order

(7)取得訂單：secure an order; build up an order

(8)取消訂單：cancel an order; cancellation of an order

(9)暫停生產：stop; withhold the production

(10)履行合約：fulfill the obligation on a contract

實例一 下試驗性訂單

Letterhead

Date: Mar. 22, 2013

Sponge Industrial Co., Ltd.

666 Queen Elizabeth Street

Brecksville, Segovia

Spain

Att: Mr.. Victor Chang/Marketing Mgr.

Dear Victor,

We are pleased to make a trial order for your sponge Model No. 1011~1020, 100 pcs per each model.

We would like you to provide them **within three weeks**. If the goods arrive on time, we shall send you **regular orders**. Please send

---

us a **P/I** for our confirmation along with promotional **literature** if available.

Looking forward to your prompt reply.

Sincerely yours,

Howard Huang

Howard Huang

❖ 譯文

我們很高興下一張貴公司品號 1011~1020 的膠棉、每款 100pcs 的試驗性訂單給貴公司。

我們希望貴公司於三週內提供這些產品。如果這些產品準時抵達，我們會定期下單。請提供一張銷貨確認書給敝公司確認，如果有相關的促銷目錄的話，請一併附上。

希望早日收到貴公司的回覆。

相關詞彙說明

within three weeks　在三週內（可用 no later than three weeks 替代）

regular orders　週期性的訂單

P/I　Proforma Invoice 的縮寫，是銷貨確認書之意，也有人譯成受訂確認書。

literature　目錄（可用 catalogue/brochure/booklet/pamphlet 替代）

實例二 追加訂單

Letterhead

Date: Mar. 22, 2013

Sponge Industrial Co., Ltd.

666 Queen Elizabeth Street

Brecksville, Segovia

Spain

Att: Mr. Jerry Ma/Marketing Mgr.

Dear Jerry, P /O :

Sub: Additional Order for our P/O#223354

Thank you for informing us that the above mentioned order is **under production**.

As it is **bull market** now, we have to increase our order. Please add 2000pcs per each items. **Apart from** this, we would like to add 10 sets of model GG-233 **at US$500 each**. Even the order has been increased, we still want you to meet the delivery terms that you offer in previous P/I. Sorry for **causing any inconvenience**.

Please confirm receipt of this order by a revised P/I. Thank you for your cooperation.

Sincerely yours,

Milton Tsai

Milton Tsai

❖ 譯文

主旨：追加訂單單號 223354 的訂量

　謝謝通知上述所提的訂單已在生產中。

　由於目前市場行情看漲，我們必須追加訂單。每款的訂購量請各增加2000個。除此之外，我們想再追加10組每組單價US$500的GG-233款式產品。即使這張訂單已追加數量，我們仍要求貴公司能在前張 P/I 所提的日期交貨。對您所造成的不便，我們甚感抱歉。

　請回傳銷貨確認書以確認已收到此份追加訂單。謝謝貴公司的合作。

相關詞彙說明

P /O　Purchase order 訂單的縮寫

under production　生產中（可用 under manufacturing 替代）

bull market　指行情看漲的市場；牛市

apart from　除此之外（可用 besides, furthermore 取代）

at US$500 each　每個單價為 US$500（若指每個產品的單價，前面的介詞是 at 或 for）

cause any inconvenience　造成任何的不便之處（是商業書信中很常用到的片語）

 因交期延遲和品質問題而取消訂單

Letterhead

Date: Mar. 22, 2013

Ericsson Corporation

536 Cutter Mill Ave

Great Neck, NY 11021

USA

Att: Mr. Vicent Lin/Marketing Mgr.

Dear Mr. Lin,

Sub: Cancell of our P/O# 12343

We are sorry to know that you will not be able to meet the requirements of **dispatching** the order by August 10. We have explained the reason why we need an urgent supply for this order in the previous letter.

In addition, we just received our P/O 12341 from you. Our **Q.C**. found out that there was not only color difference but also **out of spec**. in this delivery.

For the above two reasons, we **have no other choice but to** cancel the order. Unless you can have some improvement or solution within one month, we are forced to switch the supplier.

We look forward to hearing from you soon with regard to how you would like to **proceed**.

Sincerely yours,

Joseph Lieu

Joseph Lieu

❖ 譯文

主旨：取消訂單號 12343

我們很遺憾得知貴公司將無法如期於八月十日出貨。我方已在前封信中說明為何請貴公司緊急寄送此張訂單的原因。

此外，我們也剛收到我們訂單單號 12341 的貨物。我們品管發現此次的交貨不但貨品顏色有差異，而且還規格不符。

基於以上兩個理由，我們不得不取消這張訂單。除非貴公司能在一個月內改善或提出解決之道，否則我們將被迫更換供應商。

希望能早日收到貴公司如何處理此事的回函。

相關詞彙說明

dispatch　原意是派遣、發送、快遞。這裡指寄送、出貨之意（可用 send/hip/delivery 替代）

Q.C.　Quality Control 之縮寫，品質管制之意，簡稱品管。

out of spec.　spec. 是 specification 之縮寫，意指規格。out of 是指超出之意。out of spec. 是超出規格或不符規格之意。

Have no other choice but to + 原型動詞　沒有其他選擇只好……；不得不……

proceed　繼續進行；開始；著手；進行；展開

實例四 因貨品不符公司現行規格而無法接訂單，但推銷替代品

Letterhead

Date: Mar. 22, 2013

Ericsson Corporation.
536 Cutter Mill Ave
Great Neck, NY 11021
USA

Att: Mr. Edward Kao/Marketing Mgr.

Dear Edward,

Sub: Your P/O# EK123

Thank you very much for **granting** us the above order.

Unfortunately, the specifications and **tolerances** that you require are not yet possible with the equipment we have. **Presently**, we just purchased the needed equipment, but it will arrive only after 8 weeks. In addition, we also know that there is no other manufacturer in our region that has the equipment to meet your needs in this order.

However, we now offer several **advanced versions** which **are compatible with** what you **requested for**. Your **are encouraged to** consider our E001~E006 as alternatives. **Please be assured** that these models will meet yor present as well as future needs.

Please contact me at your convenience if you wish to **pursue** the matter in a different way.

Sincerely yours,

Benjamin Lee

Benjamin Lee

❖ 譯文

〰〰〰〰〰〰〰〰〰〰〰〰〰〰〰〰〰〰〰〰〰〰〰〰〰

主旨：貴公司訂單單號 EK123

　　謝謝貴公司授與敝公司上述所提之訂單。

　　很不幸的，以敝公司現有的機器設備而言，尚無法提供貴公司在此訂單上所要求的規格與偏差值。我們最近才購置了所需的設備，但此新設備要八週後才會入廠。此外，我們也知道我們這個領域沒有其他供應商有設備可以滿足貴公司此訂單上的需求。

　　但是，我們可以提供與貴公司需求相容的最新款式。我們促請貴公司考慮使用敝公司型號 E001~E006 的產品作為替代品。請您放心，以上之替代品將可符合貴公司現階段與未來的需求。

　　如果貴公司希望繼續以不同的方式來進行此事，請於方便時與我聯絡。

〰〰〰〰〰〰〰〰〰〰〰〰〰〰〰〰〰〰〰〰〰〰〰〰〰

相關詞彙說明

grant　給予；授與（可用 give 取代）。

tolerance　公差（指對成品的尺寸或大小等可容忍的誤差範圍）

presently　現在；目前（可用 currently 替代）。

advanced versions　先進的款式；最新款式。

be compatible with　與……共用；相容的。

request for　要求（可用 ask for 替代）。

be encouraged to　敦促；鼓勵；建議（可用 be encouraged to 替代）。

please be sured that　請放心；請安心。

pursue　繼續；進行；從事。

實例五 原料已購置進廠而無法接受取消訂單

Letterhead

Date: Mar. 22, 2013

Ericsson Corporation.

536 Cutter Mill Ave

Great Neck, NY 11021

USA

Att: Mr. Edward Kao/Marketing Mgr.

Dear Edward,

Well received your notification of cancellation of your P/O 1234. Unfortunately, this kind of notification is late for us to withhold the production.

All the materials arrived in our plant two weeks ago, and we **are about to** finish the production within three days. The **S/O** has been **booked** as well. **Along this line**, we are afraid that there is nothing can be done, except continuing the production. Otherwise, more losses will be created. If you **insist on** stopping the production, we will be forced to **charge** all the costs **at your end** for your **violating** the contract.

A quick, **concrete** reply would very much **facilitate** our procedures.

Sincerely yours,

Benjamin Lee
Benjamin Lee

❖ 譯文

　　已收到貴公司取消訂單單號 1234 之通知。很不幸的，這份通知來得太遲，已無法讓我們停止所有的生產。

　　所有的原料已於兩週前進廠，且在三天內將完成此批訂單的生產，艙位號碼也已訂定。關於這一點，除了繼續生產外，我們恐怕已不能再做什麼，不然的話會造成更多的損失。如果貴公司堅持要停止生產，我們將被迫向貴公司索取所有損失費用，因為貴公司違反了合約。

　　貴公司迅速且具體的回覆，將有助於敝公司後續的處理。

相關詞彙說明

be about to　　即將

S/O　　Shipping Order 的縮寫，艙位之意。

book　　預定；預約

along this line　　關於這一點。

be afraid　　恐怕（be afraid of+ 名詞；be afraid tht + 子句）

insist on　　堅持（in sist on + Ving）

charge　　索價；將帳記在……

charge at your end　　向貴公司索價；帳記在貴公司這一方

violate　　違反

concrete　　具體的

facilitate　　使容易；促進；幫助（可用 make easy/help 替代）

包裝、裝船與保險

Packing, Shipment and Insurance

　　包裝可以保護商品，防止在運送過程中可能造成的損壞。一般而言，包裝可區分為內包裝 (inner packing) 以及外包裝 (outer packing)，有時客戶也會依實際的需求而要求個別包裝 (unitary packing)。

　　一般而言，廠商會將商品放在內盒，再將內盒裝入出口紙箱；有時客戶會要求再將出口紙箱裝進木箱或其他特製的箱子內。有些地區已禁止使用木箱，不然會要求廠商以木箱包裝時，必須做煙燻證明 (Fumigation Certificate)；有些客戶會要求要將包裝好的貨物置於棧板 (pallet) 上，有些客戶則不要求使用棧板，只要直接疊櫃就可。疊櫃可以省材積、裝更多的貨於櫃內，想當然爾就可以節省運費，但是在裝櫃和卸櫃時比較費時和費工。使用棧板則在裝、卸櫃時，因為可使用到機器故較省時和省工，但是卻會增加材積和運費。

　　裝船對賣方而言，意味著已履行交易，並可因此取得貨款；對買方而言，意味著訂購的商品即將到手，待商品入港後，就可以憑單付款和贖貨。最基本的裝船文件包括海運提單 (B/L; Bill of Lading)、包裝單 (Packing List)、商業發票 (Commercial Invoice)。有些客戶會要求另外附上產地證明 (Certificate of Origin)、裝

船通知證明 (Shipping Advice)、裝櫃證明 (Loading Certificate)、保險單 (Insurance Policy)……等證明。在空運 (air shipment) 方面的文件包括大提單 (MAWB: Master Air Way Bill) 與小提單 (HAWB: House Air Way Bill)、包裝單 (Packing List)、商業發票 (Commercial Invoice) 等。

在運送的過程中，難免會遇到難以預測的災難，而造成莫大的損失。所以有些客戶會投保海上保險 (Marine Insurance/Sea-Water Damage Insurance) 以彌補這樣的情況發生。海上保險包括：全損險 (TLO: Total Loss Only)、單獨海損不賠／平安險 (Free from Particular Average)、水漬險 (WA: With Average)、全險 (AAR: Against All Risks)、戰爭險 (War Risk)、偷竊遺失險 (TPND: Theft, Pilferage and Non-Delivery)、罷工暴動內亂險 (SRCC: Strikes, Riots & Civil Commotion)、倉到倉險 (Warehouse to Warehouse Insurance)、確定保險 (Definite Insurance)、臨時保險 (Provisional Insurance)。所謂的臨時保險是指保險金額、目的港、發票金額、船名等有任何一項不能確定時，但又因需要而投保的險；確定保險則是指以上所提一切都已確定和清楚才是確定保險。

 包裝指示

Letterhead

Date: Mar. 22, 2013

Delta Service Incorporation

2063 N# 67th Street

Seattle WA 98115

USA

Att: Mr. Tosho Takeyama/Marketing Mgr.

Dear Takeyama san,

Sub: Packing Instructions

Well received your inquiry about the instruction. Please pack the goods according to the following instruction.

All the goods shall be packed in a **double polyethylene bag**. Put each piece of good into an **inner box**. **Lay in** 10 pcs of inner boxes into a **strong cardboard carton secured with string**. Then pack 10 cardboards in an **aluminium-foiled wooden case**. The wooden case shall be secureds with an **iron band**. Please use **hessian cloth** or **straw mat** as the **stuffing**. Please limit the weight of to 100 kgs and **metalstrap** all cases **in stacks of three** and **mark** all the case with an A in the **square**.

As for the **shipping mark**, I will revert to you in a day or two. **Thanks for your being patient**.

Sincerely yours,

Benjamin Lee

Benjamin Lee

❖ 譯文

主旨：包裝指示

已收到貴公司有關包裝指示的信函。請根據下列指示包裝貨品：

所有的貨品都應以雙層 PE 塑膠袋包裝。將每一個貨品裝入一內盒，十個內盒再裝入一個以線繩固定住的堅固外箱。然後將十個紙箱裝入內襯鋁箔之木箱內。木箱應以鐵條固定住，並以麻布或稻草團做填充。請將每箱的重量限制為 100 公斤，以三個箱子為一疊，以鐵條捆住，所有的箱子都要以正方形 A 做記號。

至於正麥的部分，我會在一兩天內回覆。謝謝貴公司耐心以待。

相關詞彙說明

double Polyethlene bag　雙層 PE 袋

inner box　內盒

lay in　放入（可用 put into 取代）

strong cardboard carton　堅固的外箱

secured with sring　以線繩固定住

aluminum-foiled wooden case　內襯鋁箔的木箱

iron band　鐵條

hessian cloth　麻布

straw mat　稻草團

stuffing　填充物

metalstrap　以鐵條捆住

in stacks of three　以三個為一疊

mark　做……的記號

square　正方形

shipping mark　正麥（指在包裝容器上標示的圖形、文字、
　　數字、字母及件號的統稱）

Thanks for your being patient.　謝謝耐心等候。

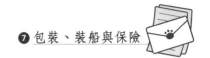

 麥頭的指示

Letterhead

Date: Mar. 22, 2013

Delta Service Incorporation

2063 N# 67th Street

Seattle WA 98115

USA

Att: Mr. Tosho Takeyama/Marketing Mgr.

Dear Takeyama san,

Sub: Packing Instructions

    Following my previous fax, I would like to offer you the following packing instructions for our P/O#12343:

    Main Marks   Side Marks

      <M.T.>   **N.W.**: 70 kgs

                                  G.W.: 72 KGS

    Item No.: 334A   **Cuft**: 5'

Description: stationary   **Msmt**: 4" × 5" × 8"

Bar code: 268787732473

CTN. No.: 1~200

Made In Japan

If there is anything remaining unclear, you are always the most welcome to contact me.

Sincerely yours,

Chiyeko Takeyama

Chiyeko Takeyama

❖ **譯文**

〰〰〰〰〰〰〰〰〰〰〰〰〰〰〰〰〰〰〰〰〰〰

主旨：包裝指示

　　承先前之傳真，提供貴公司敝公司訂單單號 12343 的包裝指示：

正麥	側麥
◇M.T.◇	淨重：N.W.: 70 kg
	毛重：72 KGS
品號：334A	材積：Cuft: 5'
描述：固定物	外箱尺寸：4" × 5" × 8"
電腦條碼：268787732473	

箱號：1~200

日本製

若有任何不清楚的地方，歡迎與我聯絡。

相關詞彙說明

following　　承……

Item No.　　產品品號

description　　產品描述

bar code　　電腦條碼；標號碼

CTN. No.　　CTN 是 carton 的縮寫，箱號的意思。

N. W.　　是 Net Weight 的縮寫，淨重之意。

G. W.　　是 Gross Weight 的縮寫，毛重之意。

Cuft.　　是 Cubic feet 的縮寫，材積之意（可用 Msmt. 替代）

Msmt　　是 Measurement 的縮寫，這裡是指外箱的尺寸。

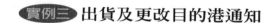

實例三 出貨及更改目的港通知

Letterhead

Date: Mar. 22, 2013

Delta Service Incorporation
2063 N# 67th Street
Seattle WA 98115
USA

Att: Mr. Adolfo Jesus/Marketing Mgr.

Dear Adolfo,

Sub: Notification of shipment & change of **destination** for P/O1971611:

Please send 600 units of this order by air freight as they are urgently in need by our customer. Please proceed with the **pending orders** by ocean freight.

Owing to the increase of **consignments** arriving at Barcelona, **discharge** of the **cargoes** seems to be much delayed. Therefore, we must ask you to send the goods to Bilbao, the north port of Spain. There should be no **alteration** in shipping date. To **cover the risk** of the incease distance, we request you **pay special attention to** packing.

Warmest personal regards,

Javier Juaquin

Javier Juaquin

❖ 譯文

主旨：訂單號 1971611 的出貨通知，以及更改貨運目的地

　　由於客戶急需，請將這張訂單的 600 個單位緊急以空運寄送，剩下的訂單數量，請以海運運送。

　　因為抵巴塞隆納港口的貨運很多，貨物卸貨的速度將會非常緩慢。所以我們必須要求貴公司將貨物送至西班牙北部畢爾包港口。出貨日的部分請不要做更動。為了避免路途增加所帶來的風險，請特別注意包裝。

　　誠摯的問候。

相關詞彙說明

destination　目的地；終點

pending orders　pending 原指懸而未決的，這裡引申為剩下的訂單。

owing to　歸因於；由於（可用 attribution to 替代）

consignment　運送；遞運品

discharge　卸貨

cargo　貨物

alteration　改變；變更（可用 change/modification 替代）

cover the risk　承擔⋯⋯的風險

pay special attention to　特別注意

warmest personal regards　誠摯的問候（不是正式的結尾語，而是用於熟人間的結尾語）

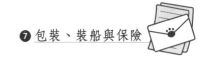

 裝船通知

Letterhead

Date: Mar. 23, 2013

Delta Service Incorporation

2063 N# 67th Street

Seattle WA 98115

USA

Att: Mr. Arthur Lai

Dear Arthur,

Sub: **Shipping Advice** for your P/O#19720709

We are glad to keep you informed that the above mentioned order was dispatched yesterday. The following are the **shipping information**:

Closing Date: 3/28

ETD: 4/2

ETA: 4/20

Vessel & Voyages: Yokohoma Hatsu V.00250-3E

Port of Destination: Valencia, Spain

We will send you the **shipping documents by courier** next week.

Best regards,

Javier Juaquin

Javier Juaquin

❖ 譯文

主旨：貴公司訂單單號 19720709 之出貨通知

　　我們很高興通知貴公司上述所提之訂單已於昨日出貨。以下是相關之出貨資料：

　　結關日：3/28

　　預計出航日：4/2

　　預計抵達日：4/20

　　船名與航次：Yokohoma Hatsu V.00250-3E

　　目的地港口：Valencia, Spain

　　敝公司會將出貨文件於下週以快遞寄出。

相關詞彙說明

shipping advice　　出貨通知；裝船通知

shipping information　　出貨資料；裝船資料

closing date　　結關日

ETD　Expected Delivery Date 的縮寫，預計開航日之意。

ETA　Expected Arrival Date 的縮寫，預計抵達日之意。

vessel & voyages　　船名與航次

port of destination　　目的地港口

shipping documents　　出貨文件；裝船文件

by courier　　以快遞

實例五　通知買方已投保完成

Letterhead

Date: Mar. 23, 2013

Delta Service Incorporation

2063 N# 67th Street

Seattle WA 98115

USA

Att: Mr. Bruce Fanshawl

Dear Arthur,

Sub: **Marine Insurance**

As per request, we have **closed the insurance** of AAR as shown in the **insurance policy** enclosed **for 100,000 on 1000 sets** of coffee machines.

The rate for AAR **insurance premium** for this order to New York is $.1.50 per $100.0. Please be advised that it is **generally adopted** by all the large **underwriters** here. Due the insurance premium for war risk is increasing **enormously**, we ask you to **cover it for your own account**.

You mentioned that you want to cover a **warehouse to warehouse insurance** for this order as well, please advise if this request is still **valid**.

Your usual prompt reply would help **a great deal** at this time.

Best regards,

Jack Peterson

Jack Peterson

❖ 譯文

主旨：海上保險

　　如貴公司所要求，敝公司已對 1000 組的咖啡機投保 100,000 的全險保險，請參考附件的保險單說明。

　　到紐約的全保險匯率是 $1.5 比 $100，此匯率乃為本地大保險商廣為採用。因戰爭險的保險費大幅度提高，我們請貴公司自行吸收這部分的費用。

　　貴公司也提到要替這張訂單投保倉庫到倉庫的保險，請告知這項請求是否還有效。

　　此次貴公司一如往常的快速回覆，將提供敝公司很大的協助。

相關詞彙說明

marine insurance　海上保險

close the insurance　完成投保手續

insurance policy　保險單

for 100,000 on 1000 sets　為 1000 組的⋯⋯投 $100,000 的保
　　險金

insurance premium　保險費

generally adopted　為⋯⋯廣泛的採用

underwriter　保險商

enormously　巨大地;龐大地

cover it for your own account　由貴公司自行承擔相關之費用

warehouse to warehouse insurance　倉到倉保險(是指由出貨
　　公司的倉庫到買方的倉庫之意)

valid　有效的

a great deal　大量的;非常

付款及逾期款項通知

Payment

　　在客戶訂單生產、交貨前或交貨後，有一個非常重要的環節就是通知客戶付款。至於是交貨前或交貨後付款，則是依據雙方所同意的付款條件來行事。

　　一般常見的付款方式有以下幾種：

訂貨付現：C.W.O.（Cash with Order）

信用狀付款：L/C (Letter of Credit)

付款交單：D/P (Documents against Payment)

承兌交單：D/A (Documents against Acceptance)

出貨前付款：Payment Before Shipment

出貨前電匯：T/T (Telegraphic Transfer) Before Shipment

郵政匯票：Postal Money Order

銀行支票：Bank Cheque

銀行本票：Cashier Cheque

記帳：O/A (Open Account)

　　在票期的部份，常見的有見票即付 (A.S. = At Sight; O.D. = On Demand)；見票後 X 天付款 (At X D/S = At X Days at Sight)，至於實際的天數則是雙方所同意的天數為條件；記帳部份有出貨後 X 天付款 (Payment After X Days from Invoiced Date)，至於實際的天數則是雙方所同意的天數為條件。

以下是與款項相關的用語：

(1)開立信用狀：open/establish/issue an L/C ; application of L/C

(2)修改信用狀：amend/modify/correct/change the L/C

(3)不可撤銷保兌信用狀：irrevocable confirmed L/C

(4)跟單信用狀：documentary L/C

(5)即期匯票：sight draft; draft at sight

(6)承兌匯票：pay / honor / settle a draft

(7)付款：effect/render/make payment

(8)付款到期日：the date of payment; the due day; the term day; the maturity

(9)累進付款：progressive payment

(10)頭期款：initial payment; down payment

(11)定期付款：punctual payment

(12)預付款：advance payment

 **通知已開立** L/C

Letterhead

Date: Mar. 25, 2013

Delta Service Incorporation

2063 N# 67th Street

Seattle WA 98115

USA

Att: Mr. Bush Clinton,

Dear Mr. Clinton,

Sub: L/C

Thank you for your shipment information. We have issued an irrevocable **confirmed L/C** today **in your favor for the amount of** US$200,000.

As usual, the **expiration** of this L/C will be on May 1st. Therefore, please make sure that the goods will be dispatched on time. The bank will **honor your draft at sight** for the amount of your

---

invoice **drawn under the L/C**. You should be able to get the original within next week.

Please confirm upon securing the L/C.

Best regards,

Timothy Copa

Timothy Copa

❖ 譯文

主旨：信用狀

謝謝貴公司之出貨通知。我們已於今天以貴公司為受益人，開立了一張金額為 US$200,000 的不可撤銷保兌信用狀。

一如往昔，此份信用狀的有效截止日期為五月一日。所以請確定產品會準時交遞。銀行會根據貴公司依信用狀所開立的發票金額來承兌即期匯票。貴公司應可於下週內收到信用狀正本。

收到信用狀後請做確認回覆。

相關詞彙說明

confirmed L/C　保兌信用狀

in your favor　以貴公司為受益人（也可用 at your favor 替代）

for the amount of　金額為……

expiration　屆滿；到期

honor your draft at sight　承兌、支付即期匯票

drawn under the L/C　以信用狀所開立的票

實例二 通知修改 L /C

Letterhead

Date: Mar. 25, 2013

Delta Service Incorporation

2063 N# 67th Street

Seattle WA 98115

USA

Att: Mr. Alfred William

Dear Mr. William,

Sub: **Amendment** of L/C#12667788

We acknowledge of receipt of the above mentioned L/C. Unfortunately, we found out that some **conditions** are incorrect. Therefore please **modify** it as following instruction:

1. As this **consignment** will be **transhipped** to Hong Kong and will be shipped separately. Please change the conditions to **partial shipment** and transhipment allowed.

2. We will not accept L/C 30 d/s. Please change it to L/C at sight.

3. In spite of our effort, we find it impossible to secure space for the shipment owing to the **unusual shortage** of **shipping space**. Please **extend** the L/C to May 15.

Thank for your cooperation. Looking forward to your prompt confirmation on above request.

Best regards,

James Chi

James Chi

❖ 譯文

〰〰〰〰〰〰〰〰〰〰〰〰〰〰〰〰〰〰〰〰

主旨：信用狀號碼 12667788 的修改

　　我們已收到上述的信用狀。很不幸的，我們發現有些條件是不正確的，所以請依據以下指示做修正：

　　1. 因為貨物將在香港做轉運，且貨物會分批運送，所以請將條款改成可允許轉運和分批運送。

　　2. 我們無法接受見票三十天後付款的條款，請將之修改為即期信用狀。

　　3. 雖然敝公司已盡最大努力，卻因為異常的船位短缺而無法保證交期，請將信用狀的有效期限延至五月十五日。

　　謝謝貴公司的合作！我們期待收到貴公司針對以上要求所做出的儘早回覆。

〰〰〰〰〰〰〰〰〰〰〰〰〰〰〰〰〰〰〰〰

相關詞彙說明

amendment　修正；更改（可用 modification/alteration 替代）

conditions　條款；條件

modify　修正；更改（可用 amend, alter, change 替代）

consignment　貨物

tranship　轉運

partial shipment　分批、部分運送

unusual shortage　異常短缺

shipping space　船位

extend　延長

實例三 通知對方付款

Letterhead

Date: Mar. 25, 2013

Delta Service Incorporation

2063 N# 67th Street

Seattle WA 98115

USA

Att: Mr. David Dickson

Dear Mr. Dickson ,

Sub: Payment for your P/O#22334

We are pleased to advise you that your P/O#22334 is already to be delivered by ocean freight . Enclosed is the commercial invoice.

Please T/T the 90% of the **invoiced amount** immediately. Once

---

we got the copy of your **remittance**, we will **effect the delivery** without any delay. Please **settle** the 10% of **balance** within one week after the goods are dispatched.

Looking forward to your prompt action.

Best regards,

Timothy Copa

Timothy Copa

❖ 譯文

主旨：貴公司訂單單號 22334 之款項

　　很高興通知貴公司訂單單號 22334 即將以海運送出。附件是商業發票。

　　請盡快先行電匯發票中 90% 款項。一收到貴公司的匯款明細影本後，我們會馬上放行貨物。請在出貨後將 10% 的餘款付清。

　　期待貴公司迅速採取行動。

相關詞彙說明

invoiced amount　發票金額

remittance　匯款（可用 T/T 或 wired payment 替代）

effect the delivery　放行

settle　結清

balance　餘款

實例四　通知支付逾期款項

Letterhead

Date: Mar. 25, 2013

Delta Service Incorporation

2063 N# 67th Street

Seattle WA 98115

USA

Att: Mr. David Dickson

Dear Mr. Dickson ,

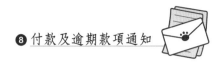

Sub: **Outstanding Payment** of US$10,000 for your P/O#22334

According to our Accounting Department records, your payment of US$10,000 for your P/O#22334 is **overdue** since March 1st.

As we have always received your payment **punctually**, we are **puzzled** to have had neither remittance nor report **in connection with** our **statement** of March 7th.

Unless we receive your payment for the due amount on your account within one week from today, we shall be forced to stop the production of your recent order.

Looking forward to your prompt action on this matter.

Best regards,

William Miller

William Miller

❖ 譯文

主旨：貴公司訂單單號 22334 US$10,000 之未付款

　　依據敝公司會計部門之記錄，貴公司訂單單號 22334，美金一萬元的款項，自三月一號起就已逾期。

　　由於敝公司總是按時收到貴公司的款項，此番未收到貴公司的電匯款，也未收到貴公司對敝公司三月七日的聲明做出回應而感到疑惑。

　　除非能在自今日算起的一週內，收到貴公司上述所提的逾期款項，否則我們將被迫停止生產貴公司最近的訂單。

　　期待貴公司對於此採取迅速的措施。

相關詞彙說明

outstanding payment　逾期款項（可用 overdue payment 替代）

overdue　逾期

punctually　按時地；如期地

puzzle　迷惑；困惑

in connection with　與……相關

statement　陳述；聲明

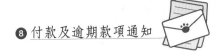

實例五 爲尚未付款致歉，並同意立即付款

Letterhead

Date: Mar. 25, 2013

Delta Service Incorporation

2063 N# 67th Street

Seattle WA 98115

USA

Att: Mr. Kevin Claudia

Dear Kevin ,

Your request regarding a $5,600 outstanding balance for our P/O11343 was immediately **checked against** our records.

Our records indicated that the $5600 had, in fact, not been paid due to **an oversight on our part**. The amount concerned was forwarded to your account of the City Bank by telegraphic transfer today. Please see the enclosed copy of our wire payment.

Please accept our **deepest apology** for any inconvenience this matter has caused you.

We look forward to the pleasure of working with you again in the very near future.

Best regards,

William Miller

William Miller

❖ 譯文

您提到的訂單單號11343，五千六百元的逾期款項一事，敝公司已迅速核對資料。

我們的記錄指出，是因敝公司的疏忽而尚未結清該筆五千六百元的款項。上述金額已於今日電匯至貴公司花旗銀行的帳號，請參考附件的電匯款影本。

請接受敝公司對此事所造成之不便的深深歉意。

我們期待在不久的將來能與貴公司合作愉快。

相關詞彙說明

check against　核對

an oversight on our part　我方的疏忽

deepest apology　深深的歉意（可用 sincere apology 替代）

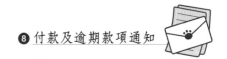

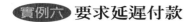 **要求延遲付款**

Letterhead

Date: Mar. 25, 2013

Delta Service Incorporation

2063 N# 67th Street

Seattle WA 98115

USA

Att: Mr. Philippe Gauchi

Dear Mr. Gauchi,

Your request regarding a $5,600 outstanding balance for our P/O 11343 has been received.

Our market has a **sharp decline** since the beginning of this season. The decline has left us with extremely heavy **burden** on our finances. We are writing today to ask for your cooperation in dealing with this problem. Specifically, we request that you grant us an additional 30 days **usance** on all payment until we feel **relief** in our finances. We have been a customer with good credit in the past years.

We have complete confidence in our ability to **ride out** this **storm**. we will be able to **make the payment** on time after 30 days.

Your positive consideration of the request would be a great help for us.

Best regards,

William Miller

William Miller

❖ 譯文

我們已收到訂單單號 11343 中五千六百元的逾期款項的要求。

從本季初開始，我們的市場就開始急劇衰退，這樣的衰退對敝公司的財務造成沉重的負擔。我們今天寫信給貴公司是希望貴公司共體時艱，尤其希望貴公司能將敝公司的欠款之付款期限延後三十天，直到敝公司的財務壓力解除為止。過去幾年，我們一直是信用良好的顧客，我們有信心能安全度過此難關。敝公司將能在三十天後付款。

貴公司能正面回覆此要求的話，將提供敝公司很大的協助。

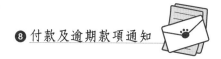

相關詞彙說明

sharp decline　急劇衰退

burden　負擔（可用 pressure 替代）

usance　票據期限

relief　減輕；緩和

ride out　安全度過

storm　原指暴風雨，在此引申為困難之意

make the payment　付款

索賠

Claim

賣方將貨物裝船，買方依約定支付款項，程序上此筆交易已經完成。可是萬一從船上卸下來的貨物和契約不符，或是發生品質不良、數量不足、產品破損或變質等情形時，就會產生索賠的問題。這時最重要的是必須先判定這樣的問題應該歸屬買方、賣方、船運公司或是運送過程中遭到自然災害所致，而後再依判定的結果做必要的賠償。許多人常將 claim（索賠）和 complaint（抱怨）混為一談，但是實際抱怨不一定與賠償有關係。所以字義上，抱怨比索賠更為廣泛。

一般而言，以下常常是造成抱怨和索賠的原因：品質不良 (inferior quality)、品質不符 (different quality)、包裝不良 (bad packing)、破損 (breakage)、短少 (shortage)、延遲交貨 (delayed shipment)、未履行契約 (breach of contract)、解約 (cancellation)、貨物毀損 (damage of the goods)、裝船文件錯誤 (wrong shipping documents)、變色 (deterioration of color) 等。

索賠常用的用語有：

(1)申請賠償：claim; claim for the damage; claim for the loss; put in a claim

(2)接受索賠：meet the claim; accept the claim

(3)處理索賠：adjust the claim; make settlement of the claim; settle the claim

(4)提出索賠：file a claim (complaint); lodge a claim (complaint); make a claim (complaint)

(5)索賠初步通知：a preliminary notice of claim

(6)向法院提出索賠訴訟：file a claim before a court; lodge a claim before a court; lay a claim before a court

(7)將索賠訴諸仲裁：submit a claim to arbitration

(8)依仲裁解決索賠：adjust the claim by arbitration

(9)合理索賠：justify a claim

(10)鑑定報告：surveyor's report

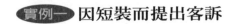

因短裝而提出客訴

Letterhead

Date: Mar. 22, 2013

S.A. Lafayette Rouge.

78 Zuai du Point du Jour-NP

101-32207 Boulogne-Billancourt

France

Att: Mr. Norbert Werner

Dear Norbert ,

Our order No. AA566 was received last week.

In checking the contents against your enclosed invoice and packing list, it was found that several items were missing. In addition, the quality is much **inferior** compared with last shipment. Please see the enclosed copy of modified invoices and packing list that reflect the items actually delivered.

Our **regulations** do not permit payment against **inaccurate** invoices. Also, some of the deleted items **render** other delivered items useless.

We await your instructions on how you **propose to** deal with this situation.

Best regards,

William Miller

William Miller

❖ **譯文**

〜〜〜〜〜〜〜〜〜〜〜〜〜〜〜〜〜〜〜〜〜〜〜〜〜〜

　　已於上週收到敝公司訂單單號 AA566 的貨物。

　　在核對貴公司所附上的發票和包裝單時，我們發現有許多產品找不到。此外，本次的貨品品質與上次出貨的品質比較後，顯得差上許多。請參考經我方修改後的實際出貨數量的發票與包裝單附件。

　　敝公司的規定是如果發票不正確，就無法付款。此外，因為有些產品的短交，致使所收的物品有些無法使用。

　　我們期待貴公司提議如何處理此狀況。

〜〜〜〜〜〜〜〜〜〜〜〜〜〜〜〜〜〜〜〜〜〜〜〜〜〜

相關詞彙說明

inferior　較差的（可用 worse than 替代）

regulations　規定

inaccurate　不正確地（可用 incorrect 替代）

render　使得；使成為

propose to　提議

實例二 因品質不良要求退貨

Letterhead

Date: Mar. 22, 2013

S.A. Lafayette Rouge.

78 Zuai du Point du Jour-NP

101-32207 Boulogne-Billancourt

France

Att: Mr. Fernando Renault

Dear Mr. Renault,

We are sorry to inform you that your last shipment for CNC machine **is not up to your usual standard**.

The goods seem to be **roughly made** and **are inclined to** be inferior. **In the interim**, we have tried unsuccessfully to make the machine **operational**. Despite the best efforts of our technical staff and your technical representatives, the product still performs far below what you **claim** in your **technical specifications**. This already cost us a great deal in wasted labor expenses and lost production.

Consequently, in view of the time already lost and the lack of any **prospect** for quick improvement and solution, we would like to return the equipment at your expense in accordance with your **guarantee**. In addition, we would like you to **be responsible for** all the expenses that caused by this inferior quality.

We look forward to your instructions on how you would like this to be **carried out**.

Best regards,

Jefferson Shakespeare
Jefferson Shakespeare

❖ **譯文**

我們很遺憾通知貴公司上次出貨的 CNC 機器並不符合貴公司一般的標準。

這些產品似乎製造粗糙且劣質。在這段期間，我們已盡力使機器正常運作，卻是一無所成。不管我們的技術人員與貴公司的技術代表多麼的努力，產品的性能還是低於貴公司技術手冊中所聲稱的水準。這種情況已造成敝公司在人力和生產成本上極大的損失。

因此，有鑑於此次的損失和缺乏改善或解決的希望，我們決定由貴公司支付費用將此機器退還，這麼做也符合貴公司的保證條款。此外，我們也希望由貴公司負責所有因不良品質而造成的相關費用。

我們期待貴公司如何處理此事的指示。

相關詞彙說明

is not up to your usual standard　不符貴公司的一般標準

roughly made　製作粗糙

be inclined to　有……的傾向

in the interim　在這段期間

operational　運作正常（可用 functional 替代）

claim　宣稱；主張

technical specifications　技術說明書；技術手冊

prospect 希望（可用 hope 替代）

guarantee 保證

be responsible for 為……負責

carry out 處理

實例三 因包裝不良導致產品損壞而索賠

Letterhead

Date: Mar. 22, 2013

S.A. Lafayette Rouge.

78 Zuai du Point du Jour-NP

101-32207 Boulogne-Billancourt

France

Att: Mr. John Brian

Dear John,

Our P/O#A88788 shipped via air fright arrived in our warehouse yesterday. Unfortunately, we found out that the goods are **severely damaged**.

To our astonishment, the goods are damaged due to bad packing. This order is scheduled to be delivered to our customer today for an **urgent installment** in a new plant. However, due to the damage, we have no choice but to **breach** the contract in delivery terms. By doing so, we will be **fined for a penalty of** US$5,000. As you did not pack the goods as per our request. We ask you to pay for all the loss caused by this accident.

We will send you all the invoices for the penalty. We are more than concerned that the **end user** will lodge this claim before a court, which will make things more **complicated**.

Your prompt action in this regard would be very much appreciated.

Best regards,

Jefferson Shakespeare

Jefferson Shakespeare

❖ 譯文

　　貴公司以空運寄出的訂單單號 A88788 已於昨日抵達敝公司倉庫。很不幸，我們發現產品受到嚴重的毀損。

　　令敝公司驚訝的是，這批貨的毀損竟然是因為包裝不良所導致。這批貨原本預計於今日送到我們的客戶手中，以便他們緊急安裝於新廠。但是，因為這樣的毀損，我們不得不違反合約上的交期要求，使我方將被處以美金五千元的罰金。因為貴公司並未依敝公司要求包裝貨物，我們要求貴公司支付所有因此意外而造成的損失。

　　我們會將罰金的發票寄給貴公司。我們更擔心的是末端使用者會向法院提出索賠訴訟，使整件事更複雜。

　　貴公司若能對此事及早採取措施將使敝公司感激不盡。

相關詞彙說明

severely damaged　遭嚴重破壞（可用 seriously destroyed 替代）

to our astonishment　令我方訝異

urgent installment　緊急安裝

breach　違反；違背（可用 violate 替代）

fine for　除以罰款

a penalty of　……的罰款

end user　終端使用者

complicated　複雜（可用 sophisticated 替代）

實例四 爲品質不良道歉並同意賠償和補貨

Letterhead

Date: Mar. 22, 2013

S.A. Lafayette Rouge.

78 Zuai du Point du Jour-NP

101-32207 Boulogne-Billancourt

France

Att: Mr. Michael Chang

Dear Mr. Chang,

　　We are terribly sorry to learn from you that the quality of goods is not satisfactory to you. We are also sorry for the **short delivery**.

　　On investigating the matter of your claim, we have noticed a **discrepancy** between our invoice and the quantities you specified. On the quantities you required, we will ship the replacement at once.

　　If you **care to dispose** of the inferior goods **at the best price obtainable**, we will sent a cheque or give you a credit note when we

---

hear from you.

In closing, let me **reiterate** our sincere regret regarding these problems. Needless to say, **your headaches are our headaches**. We are **fully committed to** preventing any **reoccurrence** of this sort of problem.

Best regards,

Jefferson Shakespeare
Jefferson Shakespeare

❖ 譯文

〜〜〜〜〜〜〜〜〜〜〜〜〜〜〜〜〜〜〜〜〜〜〜〜〜〜〜〜〜

　　我們為產品品質無法令貴公司滿意而感抱歉，也為短裝事宜向貴公司致歉。

　　在調查貴公司所申訴的事件時，我們注意到敝公司發票上的數量與貴公司所提的數量有不符之處。敝公司會針對貴公司所要求之數量，立即運送替代品。

　　如果貴公司願意以可以拿到的最好的價格處理那些不良品，我們將在接到貴公司消息後，寄支票給貴公司或給予貴公司銷貨折讓。

　　最後，請允許敝公司重申對此問題的誠摯歉意。貴公司的麻煩也就是敝公司的麻煩。我們將盡全力防止問題再度發生。

〜〜〜〜〜〜〜〜〜〜〜〜〜〜〜〜〜〜〜〜〜〜〜〜〜〜〜〜〜

相關詞彙說明

short delivery　短裝

discrepancy　不符；不一致

care to　願意；喜歡；想要

dispose　處理；處置

at the best price　以最好的價格

obtainable　能得到的

reiterate　重申；反覆做

needless to say　不用說

You headaches are our headaches.　headaches 原指頭痛之意，
　　這裡指麻煩之意。

fully committed to　全心致力於……

reoccurrence　重複發生

實例五　拒絕退貨及退款要求

Letterhead

Date: Mar. 22, 2013

S.A. Lafayette Rouge.

78 Zuai du Point du Jour-NP

101-32207 Boulogne-Billancourt

France

Att: Mr. Martine Richardson

Dear Mr. Richardson,

Your claim, outlined in your previous letter, with regard to the **malfunction** of the cutter has been given **thorough consideration** here.

Unfortunately, though we regret your **dissatisfaction** with our product, our **investigation** does not support your claim. Consequently, we are not prepared to accept the **returned** cutter and will have no alternative but to insist on payment of the **contracted amount**.

We would be happy to continue to cooperate in any way we can in facilitating the **utilization** of the cutter.

Best regards,

Jefferson Shakespeare

Jefferson Shakespeare

❖ 譯文

〜〜〜〜〜〜〜〜〜〜〜〜〜〜〜〜〜〜〜〜〜〜〜〜〜〜〜〜〜〜〜〜

貴公司在前封信提到有關切割機運作不良的客訴要求，敝公司已做過全盤的考量。

很不幸，雖然敝公司很遺憾貴公司不滿意敝公司產品，可是根據我方的調查，貴公司的索賠要求並無法成立。因此，敝公司並不準備接受貴公司將切割機退回的要求，且貴公司除了必須支付合約上的金額外別無他途。

敝公司將盡全力與貴公司合作，以確保有助於切割機的使用。

〜〜〜〜〜〜〜〜〜〜〜〜〜〜〜〜〜〜〜〜〜〜〜〜〜〜〜〜〜〜〜〜

相關詞彙說明

malfunction　運作不良（可用 bad function 替代）

thorough consideration　全盤的考量

dissatisfaction　不滿；不平

investigation　調查

return　退回；退還

contracted amount　合約金額

utilization　使用；利用

拜訪、會議、住宿、接機安排及謝函

Arrangement

　　商業書信除了與前面幾章所提的貿易流程息息相關外，其實也與其他商務安排有著密切關係。畢竟對商業人士而言，出差、會議與住宿的安排已變成生活的一部分。出差結束後，寫謝函感謝客戶的招待也已變成了例行性的公事。

　　其實出差的安排信函一點也不困難，大部分的客戶都很願意提供各方面的協助。在這樣的前提下，你的工作就只是將正確的訊息告知客戶，請客戶協助。

實例一 通知即將拜訪客戶

Letterhead

Date: Mar. 22, 2013

Address & Att.

Dear Jacob,

I will be **on a business trip** in your region during May 1st ~May 10th. I see the necessity of **paying a visit** to you for the discussion on the recent problem you are encountering with.

Please be advised will this be acceptable for you. If so, please advise what date and what time will be **the most convenient for you**. I will revert to you about the detailed **tentative schedule** and discussed agenda to you later.

As it is the **busy season** for air-line company at this moment, I would appreciate a lot if you could let me have your prompt reply soon. This will facilitate my **booking process**. Thank you.

Best regards,

Lesley Chang
Lesley Chang

❖ 譯文

　　我將於五月一日～五月十日期間到您的地區出差。我認為有必要拜訪您，以對您最近所面臨的問題做進一步討論。

　　請告知這樣的安排是否為您所接受？如果可以的話，請通知我您最方便的日期和時間。稍後我將給您預定的時間表和行程細節。

　　因為這段時間剛好是航空公司的旺季，如果您能儘早回覆，我將感激不盡。這樣將有助於我的訂位作業。謝謝！

〰〰〰〰〰〰〰〰〰〰〰〰〰〰〰〰〰〰〰〰〰〰〰

相關詞彙說明

> on a business trip　出差
>
> pay a visit　拜訪（可用 visit/call on＋人；call at＋地方）
>
> the most convenient time for you　你最方便的時間
>
> tentative schedule　預定的時間表
>
> busy season　旺季
>
> booking process　訂位作業

實例二 通知接受拜訪、議程安排

Letterhead

Date: Jul. 22, 2013

A ddress & Att.

Dear Matthew,

　　It is in deed exciting news that you will be visiting us next month. I will be here waiting for you **with** my **arms open**.

I will be available the whole day of August 5th. I can pick you up at the airport and hotel. The following are the schedule and **agenda** that I prepared. Please share with me your thought about this.

Date:	Time	Agenda
Aug. 4	???	Picking you up at the airport
		Please advise your arrival time and flight No.
Aug. 5	8:00	Picking you up at the hotel
	9:00	Arriving at my office and **warming up**
	9:30	Discussion on **exclusive** agenda
	12:00	Lunch
	13:30	Discussion on recent claim
	16:00	Send you back to the hotel for **refreshing**
	18:30	Picking you up at the hotel to a French dinner

Please let me know if you want to discuss other things else. If you need any assistance on accommodation, please let me know. **I would be glad to be of assistance for you.**

Best regards,

George Williams

George Williams

❖ 譯文

〰〰〰〰〰〰〰〰〰〰〰〰〰〰〰〰〰〰〰〰〰〰〰〰〰〰〰〰

　　聽到您將於下個月拜訪敝公司，確實是一件令人興奮的事。我將在這裡展開雙臂歡迎您的到來。

　　我八月五日整天都有空，也可以到機場和飯店接您。以下是我所準備的時間行程表和議程表。請與我分享您對此的看法。

日期	時間	議程
Aug. 4	???	到機場接您
		請通知抵達時間與班機號碼
Aug. 5	8:00	到飯店接您
	9:00	抵達我方公司並做準備
	9:30	討論獨家代理的議題
	12:00	午餐
	13:30	討論最近的客訴
	16:00	送您回飯店休息恢復精神
	18:30	到飯店接您享用法式晚餐

　　請讓我知道您是否還要討論其他的議題。如果您需要住宿方面的協助，請讓我知道。我很樂意協助您。

〰〰〰〰〰〰〰〰〰〰〰〰〰〰〰〰〰〰〰〰〰〰〰〰〰〰〰〰

相關詞彙說明

with arms open　展開雙臂

agenda　議程

warm up　做準備

excluisive　獨家的

refresh　補充精神；恢復精神

I would be glad to be of assistance for you.　我很樂意為您提供協助。（是一句書信中常用的客氣用語或結尾語）。

實例三 通知班機抵達時間並請求代訂飯店

Letterhead

Date: Mar. 22, 2013

A ddress & Att.

Dear Thompson,

Thank you for confirming that you will be available on April 5th.

The following are my **flight schedule**. No need to pick me up at the airport. I can **take a shuttle bus** to my hotel on my own.

Date:	Departure	Arrival	From	Destination	Flight No.
Apr. 2	05:00	24:00	Taipei	Barcelona	CX556
Apr. 4	10:30	14:30	Barcelona	Paris	Lux533

However, I do need your **assistance on making hotel reservation for** me at SOFITEL for the night of April 2 ~3. A standard room will be fine with me.

Thank you for all your assistance. I can't wait to see you now.

Best regards,

Kyle Kirks

Kyle Kirks

❖ 譯文

～～～～～～～～～～～～～～～～～～～～～～～～～～～～～

謝謝您告知您四月五日那天有空。

以下是我的班機時間。您不需要到機場接我，因為我可以自行坐接駁巴士到下榻的旅館。

日期	出發時間	抵達時間	出發地	目的地	班機號碼
Apr. 2	05:00	24:00	台北	巴塞隆納	CX556
Apr. 4	10:30	14:30	巴塞隆納	巴黎	Lux533

然而，我需要您幫我預約 SOFITEL 飯店四月二號和三號兩個晚上的房間。訂一間標準房就可以了。

非常謝謝您的協助，我等不及要與您碰面了。

相關詞彙說明

flight schedule　班機時間表

take a shuttle bus　搭接駁巴士

assistance on　在⋯⋯的協助

making a hotel reservation for　為⋯⋯預約飯店

實例四 通知更改會面日期

Letterhead

Date: Mar. 28, 2013

A ddress & Att.

Dear Mark Bush,

I am glad to hear that you will be in Spain from April 5 to 8.

Unfortunately, I will be in Munich for **attending a conference** at the same dates. This is **a long-standing commitment** made early this year. Consequently, I will not be able to meet you then.

However, I would be very much like to **get to see you** this trip since there are several important matters to discuss. Though I realize how difficult it is to **adjust schedules on short time notice**, I would appreciate it if you could somehow arrange to meet me **some day** before April 5 or after April 8. Otherwise, I have no other choice, but arrange Mr. Morrano Gonzale, our managing director, to meet with you. He will be ready to **engage in** discussion with you and **take the necessary action**.

Please let me know if this is possible.

Best regards,

George Williams

George Williams

❖ 譯文

很高興得知您將於四月五日～八日停留西班牙。

很不幸，我將在同樣的日期前往慕尼黑參加會議。這個會議早在今年年初就排定了。因此，我無法於那段時間與你碰面。

但是，真的很想趁您此趟行程與您碰面，因為有許多重要的議題尚待討論。雖然在這麼短的時間內通知您，勢必造成您行程重新調整的諸多困擾。如果您可以安排在四月五日前或四月八日後安排與我碰面，我將感激不盡。不然我只好請敝公司的總經理 Morrano Gonzale 先生與您碰面。他將致力於與您做相關的討論，並在必要的時候採取任何決定或措施。

請讓我知道以上的安排是否可行。

相關詞彙說明

attend a conference　參加會議（可用 attend a meeting; partic-
　　ipate in a meeting 替代）

a long-standing commitment　很久以前約定好的事

get to see you　見到你（是 see you/meet you 的口語用法）

adjust schedule　調整時間表

on short time notice　在很短的時間內做通知（可用 notifica-
　　tion 取代 notice）

some day　特定的一天（與 someday 不特定的某一天不同。
　　some day 是表示等待被確定的特定一天，而 someday
　　則是不確定，不特定的某一天）

engage in　忙於；從事；埋頭致力於

take the necessary action　採取必要的行動或措施

實例五 為出差時所受到的照顧致謝

Letterhead

Date: Mar. 28, 2013

A ddress & Att.

Dear Salvador,

I am writing you to thank you for the time you spent with me and also for the **hospitality** you **extended** my recent visit to your country.

The **delightful and memorable** lunch at your **headquarters** brought to me again the close relationship between our two companies. Thank you as well for the beautiful vase you gave me at the welcome party.

I look forward to the chance to **reciprocate** on the **occasion** of your next visit to my country. I also look forward to developing that relationship to the **mutual benefit** for our organizations.

Best regards,

Paco Jaimei
Paco Jaimei

❖ 譯文

　　對您在我此次拜訪貴國時，您所給予的時間和熱情招待等種種的一切，在此向您致謝。

　　在貴公司總部所享用的開心且難忘的午餐使我們兩家公司的關係更為密切。也非常謝謝您在歡迎茶會中所送的漂亮花瓶。

　　我期待在您下次拜訪敝公司時能回報您。我也期待與貴公司發展雙方互惠的關係。

相關詞彙說明

hospitality　熱情招待（可用 warmest reception 替代）

extend　給予；提供

delightful and memorable　令人愉快和難忘的

headquarters　總部；總公司

reciprocate　報答；回報

occasion　場合；時機（可用 opportunity/chance 替代）

mutual benefit　互惠（可用 mutual profit/mutually profitable 或 reciprocal profit/benefit 替代）

國家圖書館出版品預行編目資料

英文商業書信一看就會 / 我的外語編輯小組編著.
-- 初版 .-- [臺南市]：我的文化 , 2012.11
面；公分

ISBN 978-986-88888-8-3（平裝）

1. 商業書信 2. 商業英文 3. 商業應用文

493.6   101021989

英文商業書信一看就會

編　　著／我的外語編輯小組

總 編 輯／林淑貞　美術主任／郭高溢

執行編輯／林怡安　封面設計／盧翊嘉

出 版 者／我的文化出版有限公司

World Cultural Publishing Company

總代理發行／世一文化事業股份有限公司

地　　址／臺南市新樂路 46 號

電　　話／(06)2618468（代表號）

傳　　真／(06)2646349

郵　　撥／03880637（國內郵購金額若不足 500 元，
　　　　　　　　　　　請加附掛號郵資 60 元）

印　　刷／世一文化事業股份有限公司

歡迎來電親洽。電話：06-2618468 # 2550

E – MAIL ／ acme00@ms4.hinet.net

2013 年 1 月初版